Oxytocin and Social Function
http://dx.doi.org/10.5772/intechopen.107645
Edited by Wei Wu

Contributors
Olih Solihin, Ahmad Zakki Abdullah, Kurniati Devi Purnamasari, Maulana Irfan, Yasundari Yasundari, Yuni Mogot, Guangxin Yue, Amit Bhalla, Sandeep Kaushal, Diego Armando Leon-Rodríguez, Luis Fernando Cardenas, Julian Becerra, Juan Carlos Caicedo Mera, Jorge Martínez Cotrina, Diego Mauricio Aponte Canencio, Yuji Yamada, Tomiki Sumiyoshi, Naira G. Hunanyan, Knarik V Kazaryan, Tatevik A. Piliposyan, Cristina Velazquez-Marrero, José R. Lemos, Thendo Gertie Makhado, Lufuno Makhado

First published in London, United Kingdom, 2024 by IntechOpen
IntechOpen is the global imprint of INTECHOPEN LIMITED, registered in England and Wales, registration number: 11086078, 5 Princes Gate Court, London, SW7 2QJ, United Kingdom

British Library Cataloguing-in-Publication Data
A catalogue record for this book is available from the British Library

Additional hard and PDF copies can be obtained from orders@intechopen.com

Oxytocin and Social Function
Edited by Wei Wu
p. cm.
Print ISBN 978-1-83769-117-3
Online ISBN 978-1-83769-118-0
eBook (PDF) ISBN 978-1-83769-119-7

For EU product safety concerns:
IN TECH d.o.o., Prolaz Marije Krucifikse Kozulić 3, 51000 Rijeka, Croatia,
info@intechopen.com or visit our website at intechopen.com.

Meet the editor

Dr. Wei Wu is a professor and associate department chair in the Department of Toxicology, Nanjing Medical University, China, where he received his Ph.D. in Toxicology. He was a guest researcher at the National Institute of Environmental Health Sciences (NIEHS) between 2017 and 2018. Dr. Wu is a member of different national and international societies in the fields of human reproduction and toxicology and has received awards from many national societies for the originality and quality of his projects. Dr. Wu has authored eighty-four peer-reviewed papers in international journals. He has edited five books and collaborated on ten books, nineteen patents, and in the organization of four international conferences. He is a reviewer of 108 journals.

Contents

Preface

As a hormone, oxytocin is closely related to human social behavior, and it has been widely studied in the fields of biology, psychology, and sociology. In recent years, with the development of research on human social behavior, the relationship between oxytocin and social function has also attracted more and more attention. This book provides a comprehensive introduction to the research on oxytocin and social function, including the relationship between oxytocin and human social behavior, the mechanism of oxytocin in regulating social function, and the application of oxytocin in improving social function. The book includes contributions from leading experts in the field of oxytocin research, as well as relevant research results and findings. This book will be of interest to researchers in biology, psychology, and sociology, as well as to people who are interested in human social behavior and want to learn more about the role of oxytocin in social function. We hope that this book will help readers deepen their understanding of oxytocin and its role in human social behavior, promote the research and application of oxytocin in the field of social function, and contribute to improving human social behavior and social development.

Chapter 1 discusses the role of depolarization-induced oxytocin vs. vasopressin secretion in the absence of external calcium, and calcium release from ryanodine-sensitive internal stores as a significant physiological contributor to neuropeptide secretion from hypothalamic neurohypophysial system terminals. This has important therapeutic implications given that exogenous administration of oxytocin to children with autistic spectrum disorders has shown some success in improving social behavior and lessening anxiety.

Chapter 2 outlines the current knowledge of oxytocin and epilepsy, including the potential mechanisms of oxytocin's antiepileptic effects, the limitations and challenges of clinical studies, and future research directions and implications. The chapter also discusses the broader impact of oxytocin research on understanding social behavior and neurological disorders.

Chapter 3 overviews studies of transcranial direct current stimulation on social cognition and discusses optimal brain regions to be targeted for ameliorating symptoms and cognitive disturbances of schizophrenia.

Chapter 4 discusses the regulation of oxytocin on empathy and its neural mechanism. Empathy plays a vital role in social communication, and it is very important for establishing harmonious relationships, trust, and mutual understanding. Oxytocin can enhance emotional and cognitive empathy, as well as trust and cooperative behavior.

Chapter 5 discusses the oxytocin receptor gene polymorphisms and event-related potentials in humans. It has been found that genetic variations of the oxytocin receptor significantly influence neural activity related to emotional and social processing, except for the early phases of face recognition.

Chapter 6 discusses oxytocin and its congeners in obstetrics practice. Carbetocin, an analogue of oxytocin, has more pronounced pharmacological effects. Heat-stable carbetocin is a promising alternative to oxytocin.

Chapter 7 reviews the social and behavior change communication framework in addressing stunting. This framework integrates principles of communication theory and social psychology to create more effective messages for behavior change. The social and behavior change communication framework can help increase public awareness of health issues, motivate them to change unhealthy behaviors, and encourage healthier behavior.

Chapter 8 discusses the morphofunctional characteristics of rythmogenic regions in the rat myometrium, as well as the identification of driver pacemaker areas under the influence of oxytocin.

The completion of this book was supported by the National Key R&D Program of China (2022YFC2702900), National Natural Science Foundation of China (82273662). Many individuals made this book a reality. The completion of this book would not have been possible without the efforts of numerous contributors. I would like to thank Mrs. Maja Bozicevic at IntechOpen for her strong support from the inception to the completion of this book. I would also like to acknowledge my coauthors for their efforts.

Wei Wu
Department of Toxicology,
School of Public Health,
Nanjing Medical University,
Nanjing, China

Chapter 1

Modulation of Oxytocin Release by Internal Calcium Stores

Cristina Velázquez-Marrero and José R. Lemos

Abstract

This chapter elucidates the role of depolarization-induced oxytocin (OT) vs. arginine vasopressin (AVP) secretion in the absence of external calcium, and calcium release from ryanodine-sensitive internal stores as a significant physiological contributor to neuropeptide secretion from hypothalamic neurohypophysial system (HNS) terminals. This has important therapeutic implications, given that exogenous administration of OT to children with autism spectrum disorders (ASD) has shown some success in improving social behavior and lowering anxiety. However, this nonspecific treatment has side effects, including seizures, increased heart rate variability, and psychotic symptoms. Alternatively, facilitating the physiological neuronal release of OT but not AVP from the HNS via modulation of ryanodine vs. inositol triphosphate receptor (IP_3R) calcium stores would specifically facilitate central vs. peripheral OT release in ASD patients.

Keywords: oxytocin, vasopressin, neurohypophysis, ryanodine receptor, inositol trisphosphate receptor, autism spectrum disorders, ADP-ribosyl cyclase/CD38

1. Introduction

1.1 Hypothalamic neurohypophysial system (HNS)

The main mechanism for neuropeptide release from neurohypophysial terminals (NHT) acts via depolarization-secretion coupling. This refers to the relationship between neuronal depolarization and the subsequent release of hormones, specifically oxytocin (OT) and arginine vasopressin (AVP), from the posterior pituitary gland (**Figure 1**). This hypothalamic-neurohypophysial system (HNS) plays a critical role in regulating various physiological processes, including social behavior, reproduction, and water balance. Within the HNS, specialized neurons located in the hypothalamus synthesize and package OT and AVP into vesicles. These magnocellular neurons (MCNs) extend their axons through the pituitary stalk and terminate in the posterior pituitary (also known as (aka) neurohypophysis) gland, where the hormones are stored and released into the capillary bed for systemic delivery.

The classic understanding of neuropeptide release involves depolarization of the hypothalamic neurons which receive excitatory input which activates action potentials to their terminals. This depolarization leads to the opening of voltage-gated

IntechOpen

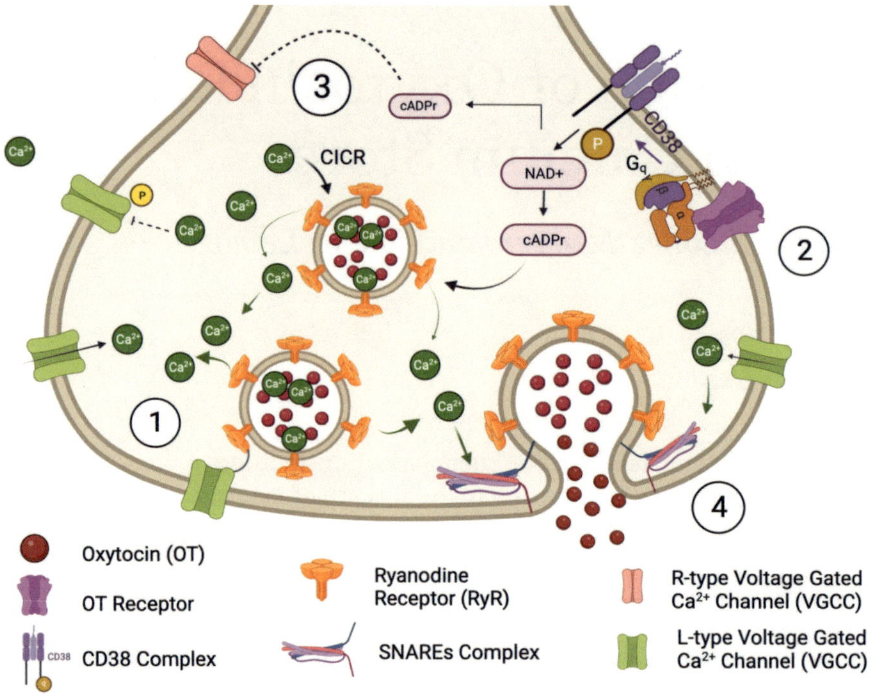

Figure 1.
Mechanisms of $[Ca^{2+}]_i$ affecting oxytocin (OT) release. 1) depolarization of L-type voltage-gated calcium channel (VGCC) mechanically opens the ryanodine receptor (RyR), which leads to the release of calcium from ryanodine-sensitive stores. 2) activation of the OT receptor (OTR) initiates cyclic adenosine diphosphate ribose (cADPr) signaling, presumably via Gq activation of ADP-ribosyl cyclase/CD38 (cluster of differentiation 38) complex catalyzing the conversion of NAD^+ into cADPr. Cyclic ADPr (cADPr) subsequently leads to the activation of ryanodine receptors (RyRs) on neurosecretory granules (NSGs) in the terminals. 3) inhibition of R-type VGCC. The release of diffusible second messenger Ca^{2+} from ryanodine-sensitive stores subsequently results in 4) OT release via SNARE complex activation. Notably, CD38, found only in OT terminals, facilitates hormonal secretion by releasing intraterminal Ca^{2+} and Ca^{2+}-induced Ca^{2+} release (CICR), other RyR types, and/or other stores [1]. This would traffic more vesicles to release sites, thus, facilitating release.

calcium channels on the terminal membrane, allowing calcium ions (Ca^{2+}) to enter (**Figure 1**). The influx of calcium into the neurons initiates a cascade of events that ultimately lead to the exocytosis of OT- and AVP-containing vesicles into the bloodstream. The increase in intracellular calcium concentration serves as a key signal for the fusion of vesicles with the neuronal membrane, resulting in the secretion of hormones into the extracellular space. The coupling between depolarization and hormone secretion involves several important steps. When calcium enters the neurons, it binds to proteins such as synaptotagmin, which triggers the fusion of hormone-containing vesicles with the plasma membrane. This fusion process allows the release of OT and AVP. Additionally, calcium influx also activates calcium-dependent enzymes, such as protein kinases, which can modulate the secretion process. These enzymes can regulate the activity of proteins involved in vesicle fusion and neurotransmitter release, thereby fine-tuning the coupling of depolarization and hormone secretion.

In this chapter, we highlight the coupling of depolarization and secretion in the HNS via the release of calcium (Ca^{2+}) from intraterminal Ca^{2+}, specifically voltage-and ryanodine-sensitive, stores as opposed to Ca^{2+} influx via voltage-gated calcium channels (VGCCs). In the absence of external Ca^{2+}, depolarization-induced secretion of oxytocin

(OT) and arginine vasopressin (AVP) can still occur, albeit with certain limitations. Recent evidence supports the hypothesis that Ca^{2+} from internal stores plays a crucial role in triggering the exocytosis of hormone-containing vesicles from nerve terminals. These alternative mechanisms can contribute to the release of OT and AVP in the absence of external calcium, thereby helping to shape both the amount and frequency of neuropeptide release differentially from OT vs. AVP terminals.

1.2 Calcium stores

These internal calcium stores within the HNS have originally referred to the endoplasmic reticulum (ER), which is a specialized organelle within cells involved in calcium storage and release. When an action potential reaches the neurohypophysial terminal, voltage-gated calcium channels on the plasma membrane open, allowing calcium ions (Ca^{2+}) to enter the terminal from the extracellular space. This calcium influx triggers the release of calcium from the internal stores, specifically the ER and the granules containing OT and AVP themselves, through a process known as calcium-induced calcium release (CICR). The released calcium from intraterminal stores amplifies the calcium signal and contributes to the overall calcium concentration in the cytoplasm. This increased calcium concentration is crucial for the fusion of neurotransmitter-containing vesicles with the plasma membrane, facilitating the release of oxytocin and arginine vasopressin into the bloodstream.

Within neurohypophysial terminals (NHT), a distinct intraterminal calcium store is characterized by local, voltage-dependent Ca^{2+} transients, known as syntillas [2, 3]. Syntillas are unaffected by the removal of extracellular Ca^{2+}, are mediated by ryanodine receptors (RyRs) within terminals, and are increased in frequency, in the absence of extracellular Ca^{2+}, by physiological levels of depolarization. The physiological role of these syntillas is under continued investigation, with recent important findings adding to the unique mosaic of regulation for OT and AVP release.

1.3 Social behavior and OT

Autism spectrum disorders (ASDs) are characterized by defects in reciprocal social interaction and communication and occur either sporadically or in a familial pattern [4–6]. The etiology of ASDs remains largely unknown and pharmacological treatments are needed. Oxytocins role in social memory and behavior, communication, and emotional recognition has now been well established [7–10]. Although making inferences about central OT functioning from peripheral measurement is difficult, the data suggest that OT abnormalities may exist in autism and that a more direct investigation of central nervous system's (CNS's) OT function is warranted [11].

2. Intraterminal $[Ca^{2+}]_i$ stores

When oxytocin neurons in the hypothalamus are activated, the depolarization of these neurons leads to the opening of voltage-gated calcium channels (VGCCs) followed by a Ca^{2+} influx through the plasma membrane. This influx of calcium ions triggers a series of events that result in the release of oxytocin both from the dendrites surrounding the cell bodies and the axonal terminals located in the neurohypophysis/posterior pituitary. However, internal calcium stores play a crucial role in maintaining calcium homeostasis and regulating calcium signaling within the oxytocin HNS neurons. The ER is responsible for sequestering and storing calcium ions, which are released upon feedback stimulation in the cell bodies via activation of IP_3R (inositol triphosphate receptors). Less is known about the accumulation of Ca^{2+} in the neurohypophysial granules themselves, which, when released, play a significant role in modulating OT secretion from terminals. Research has shown that an increase in intracellular calcium concentration acts as a signal for the secretory granules containing oxytocin to undergo "priming" characterized by the mobilization of granules from the releasable pool to the readily releasable and subsequently, the immediately releasable pool.

Once granules are trafficked to the immediately releasable pool, calcium influx via VGCC binds to specific proteins, such as the synaptotagmin/SNARE complex, which facilitates the fusion of the secretory granules with the plasma membrane (**Figure 1**). This fusion allows the release of oxytocin into the extracellular space, where it can act on target tissues and receptors. Calcium stores within terminals are activated by three known mechanisms: calcium-induced calcium release (CICR), depolarization via a direct link of ryanodine receptor (RyR) and L-type VGCC, and ligand-mediated G-protein receptors leading to the activation of specific signaling cascades. CICR is an essential step in the amplification of the calcium signal eliciting further release from IP_3 and RyR-mediated internal stores while inactivating specific VGCC. Activation of the IP_3R and RyR via calcium or by second messengers, such as IP_3 or cyclic adenosine diphosphate ribose (cADPr), leads to the release of further calcium into the cytoplasm. Therefore, while calcium is key to triggering the process of secretion, the source of calcium necessary for fine-tuned depolarization-induced OT release comes from multiple sources that interact to optimize release.

2.1 Original findings

Previously, in experiments monitoring not only $[Ca^{2+}]$ but also AVP release from HNS terminals in response to K^+ depolarization, it was shown that none of the classical modulators of intracellular Ca^{2+} release, such as caffeine, affected any of these measurements [12]. In contrast, several population (using dissociated NHTs) release experiments have been performed [1, 2, 13, 14], assessing the effects of both ryanodine and caffeine on basal neuropeptide release. Caffeine, a strong agonist of RyRs [15], induces an increase in neuropeptide release from these terminals, even in the absence of extracellular Ca^{2+} [1, 13, 14]. Ryanodine, at a concentration at which ryanodine is known to inactivate RyRs [15], has been shown to half-block this caffeine-induced release [14, 16]. This is similar to its effects on Ca^{2+} spark (syntilla) frequency in these NHTs [2]. Our results indicate that depolarization-induced neuropeptide secretion is present in the absence of external calcium, and calcium release from ryanodine-sensitive internal stores is a significant physiological contributor to neuropeptide secretion from HNS terminals.

2.2 Syntillas and calcium stores within NT

More conclusive evidence for Ca^{2+} stores in NHTs came from our studies demonstrating the presence of spontaneous focal Ca^{2+} transients in mouse neurohypophysial terminals [2, 17]. Since these Ca^{2+} syntillas were found in the absence of extracellular Ca^{2+}, they had to arise from intracellular stores. Additionally, the rate of syntillas is affected by agonists/antagonists of the RyR, a channel that is well known to control the release of [Ca^{2+}]. Dihydropyridine receptors (DHPRs) function as voltage sensors within terminals of MCNs. The DHPRs appear to be linked to type-1 RyRs in a manner bearing similarities to the mechanism in skeletal muscle [17, 18]. Data from immunocytochemistry, Western blot analysis, and electrophysiology demonstrate the existence of type-1 RyRs linking neuronal activity, as signaled by depolarization of the plasma membrane, to a rise in [Ca^{2+}] in nerve terminals [12]. These results support previous findings [17] indicating the role of type-1 RyRs in response to depolarization and imply its possible physiological significance in depolarization secretion coupling (DSC) (**Figure 1**).

The cyclic ADP-ribose (cADPr) signaling pathway (**Figure 1**) initiates a signaling cascade leading to activation of the RyRs in vivo and subsequent release of [Ca^{2+}] from ryanodine-sensitive stores [19–21]. Interestingly, in NHTs, blocking cADPr signaling was shown to attenuate high [K^{+}] -induced rises in [Ca^{2+}] but only inhibited OT release from isolated terminals [22, 23]. This strongly suggests that the cADPr pathway is present in OT terminals [1] and linked to neuropeptide release (**Figure 1**). Cyclic ADP ribose hydrolase (CD38) is a catalyst for the formation of cADPr and nicotinic acid adenine dinucleotide phosphate (NAADP) by ADP-ribosyl cyclase from nicotinamide adenine dinucleotide (NAD) and NAD phosphate [24]. Both are known to release Ca^{2+} from intracellular ryanodine-sensitive pools as part of a second messenger-signaling pathway.

2.3 NH terminals compared to HNS cells

Facilitation of neuropeptide release from magnocellular neuron (MCN) somatodendrites can be induced by increased [Ca^{2+}] [25]. This facilitation appears to be due to the trafficking of neurosecretory granules (NSGs) toward the cell surface [26]. Such facilitation increases the somatodendritic secretory response to signals, such as OT, that mobilize intracellular Ca^{2+} (via IP_3 receptors) in ER [25]. AVP and OT normally elicit little somatodendritic secretion, but after facilitation, each neuropeptide elicits enhanced somatodendritic secretion from their respective MCNs [25, 27]. This includes the response to somatodendritic action potentials, which typically do not result in local release, but can alter facilitation by AVP or OT.

3. AVP vs. OT in terminals and HNS cell bodies

3.1 Differences in VGCC

There are subtle differences between the two types of nerve terminals concerning the distribution of the distinct types of VGCCs. The AVP terminals show the L-, N-, and P/Q types of VGCCs, while the OT terminals have the L-, N-, and R types [13]. In contrast, the [Ca^{2+}] responses induced by high K^{+} (50 mM) involve all Ca^{2+}-channel subtypes in both AVP and OT somata [28]. Electrophysiological studies on the Ca^{2+} channels in the NH nerve terminals have shown that specific types (Q vs. R) of Ca^{2+} channels are uniquely important in AVP versus OT secretion (**Figure 1**).

3.2 Primed release of OT

Mechanisms involved in neurosecretion from the nerve terminals of the AVP- and OT-containing neurons of the HNS are different between this compartment and their somatodendritic region. OT has positive and AVP has various (positive, negative, or no effect) feedback on their release from somata and dendrites, but not from NHTs. Instead, voltage is the primary regulator in terminals. Both compartments utilize intracellular release of Ca^{2+} to regulate the release of OT but not AVP, which appears regulated primarily by Ca^{2+} entry through VGCCs. However, MCN dendrites utilize IP_3Rs in ER, while NHTs utilize RyRs in NSGs to regulate OT release. Nevertheless, the trafficking of NSGs is the main mechanism for the facilitation of release in both compartments [13, 25, 29]. SNARE-mediated exocytosis is also different in somatodendritic versus NHTs. Thus, these HNS compartments are different in their regulatory mechanisms for neurosecretion. Since OT and AVP are also found in CNS axons and terminals, future experiments are required to determine if differences in the central release (see **Figure 2**) of these peptides compared to the NHTs or somatodendritic regions will have subsequent consequences on their behavioral effects.

3.3 Physiological relevance (bursting, behavior, etc.)

Magnocellular neurons in the supraoptic (SON) and paraventricular (PVN) nuclei (see **Figure 2**) exhibit bursting patterns leading to secretion of oxytocin in a characteristic episodic or pulsatile manner rather than in a continuous or steady stream [13, 32, 33]. The bursting pattern is regulated by various physiological and environmental factors, such as sensory stimuli, including touch, suckling, and sexual activity. These stimuli activate specific brain regions, leading to increased activity in oxytocin

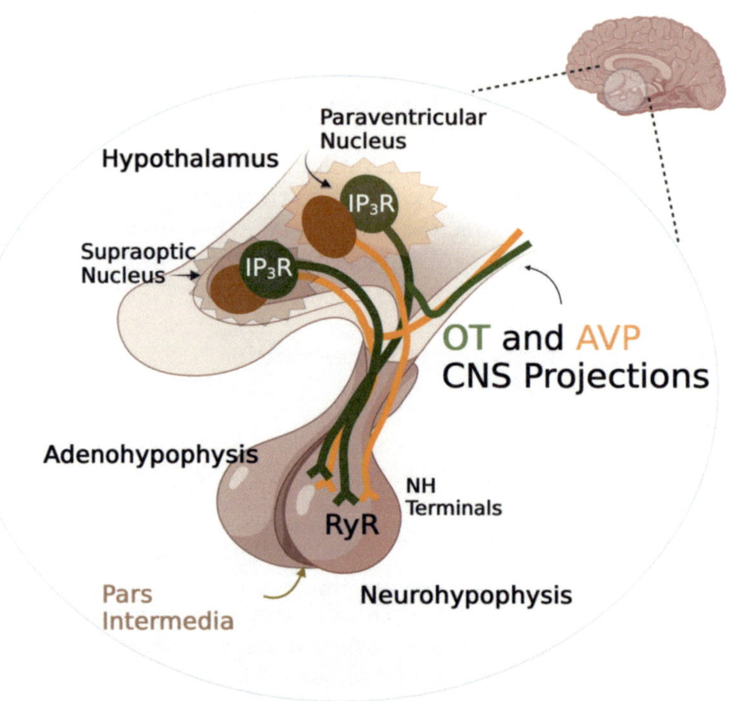

Figure 2.
The diagram of hypothalamic neurohypophysial system (HNS) showing cell bodies in hypothalamic supraoptic (SON) and paraventricular nuclei (PVN) projecting to terminals in neurohypophysis and in central nervous system (CNS) [22, 30, 31]. The priming of release takes place via IP_3R (inositol triphosphate (IP_3) receptors) in oxytocin (OT) cell bodies but via RyR (ryanodine receptors) in OT terminals.

neurons and subsequent bursts of oxytocin release [34]. The bursting pattern of oxytocin secretion is particularly relevant during reproductive processes, including childbirth, breastfeeding, and bonding between individuals [35–37].

The episodic nature of oxytocin release allows for precise and selective activation of oxytocin receptors, ensuring proper physiological responses. The precision of these responses highlights the importance of linking OT release with appropriate physiological and environmental signaling. The pattern of the electrical activity of OT neurons exhibits a sustained outwardly rectifying potential, as well as a consequent depolarizing rebound potential, not found in AVP neurons [38]. OT neurons further exhibit specific voltage-gated calcium channels, R-type, which are modulated by distinct pathways not associated with AVP neurons [39, 40]. Furthermore, in OT neurons, but not in AVP, the activity and modulation of release are highly dependent on the presence of the second messenger, phosphatidylinositol 4,5-bisphosphate (PI(4,5)P2) [41], which permissively gates N-type channels that contribute to the Ca^{2+} influx during spike trains [42] and mediates positive feedback via activation of OT receptors (OTRs) (**Figure 2**) within the dendritic arborization of cell bodies in SON and PVN [43, 44].

The oxytocin receptor and inositol trisphosphate (IP_3) are interconnected in a signaling pathway that mediates the effects of oxytocin on cellular responses. When oxytocin binds to its receptor, it leads to the activation of a G protein associated with the receptor. This activation causes the G protein to exchange GDP (guanosine diphosphate) for GTP (guanosine triphosphate), leading to the dissociation of the G protein into its α, β, and γ subunits [45]. The α subunit of the G protein then activates phospholipase C (PLC) activating a cascade of intracellular events, including the generation of IP_3. The IP_3 messenger acts as a diffusible molecule that binds to IP_3 receptors (IP_3Rs) located on the endoplasmic reticulum (ER) membrane in OT neuronal cell bodies releasing calcium ions (Ca^{2+}) from the ER stores into the cytoplasm. This increase in intracellular calcium concentration triggers various cellular responses, including modulation of synaptic transmission in the brain. Notably, so far, the presence of endoplasmic reticulum and subsequent IP_3 signaling has not been observed in neurohypophysial terminals. It is worth noting that the interactions between oxytocin receptor signaling and IP_3 are just one aspect of the broader mechanisms underlying intercellular calcium stores and oxytocin's physiological effects. Having shown that DSC occurs in the absence of extracellular Ca^{2+} and is independent of VGCCs lends support to the general premise of the Ca^{2+}-voltage hypothesis of Parnas and colleagues [46]. Whether this process modulates or plays a key role in the initiation and/or the termination of physiological release during a burst of action potentials in the HNS, however, remains to be proven.

4. Potential therapeutic approaches for treating autism

4.1 Current understanding of the role of OT

Autism spectrum disorder (ASD) encompasses multiple complex and behaviorally defined disorders characterized by impairments in social interaction, memory, communication and language, repetitive behaviors, as well as in the range of interests and activities shown by autistic patients [47–52]. Genetic [53] and environmental factors [49, 53–57] have been identified in the development of ASD. For example, the use of the anticonvulsant drug valproate during pregnancy increases the risk for the development of autism in children. Elevated OT levels within various brain networks have been associated with enhanced social cognition and emotional recognition [9]. A strong OT receptor labeling pattern has been observed in the medial prefrontal cortex, ventral-tegmental area, limbic and prelimbic areas [58], which are brain areas associated with social bonding, cognition, emotion, and reward. Conversely, low OT levels and a downregulated OTR gene are found in people diagnosed with various forms of ASD [59–61].

Conversely, Williams syndrome is characterized by abnormal social behavior (characterized by reduced social inhibition, an increased affinity toward attending to faces, and reduced sensitivity to fear-related social stimuli) and increased oxytocin functioning [62]. Furthermore, an empirical research study [63] demonstrated that the OT receptor is overexpressed in Williams syndrome.

4.2 Treatments currently in use and their prognosis

Importantly, OT and OT receptor knockout (KO) mice exhibit social impairments similar to those associated with ASD [58], and administration of OT to these KO mice can "rescue" them from ASD-like behaviors [8], showing that such therapy is possible. Similarly, a defective OT release process has been found to mediate the drop in OT levels in another autistic animal model, CD38 KOs [22, 23]. But, so far, treatments (as with OT injections or sprays) have focused on systemic OT levels. For example, Higashida et al. [22] have used agents (all-trans-retinoic acid) that target CD38 in the neurohypophysis. We hypothesize that instead treatments should target brain regions such as magnocellular somata (**Figure 2**) and their central nervous system (CNS) projections (e.g., medial prefrontal cortex, ventral-tegmental area, limbic and prelimbic areas) [58]. This could be done by agents that affect IP_3Rs instead of RyRs (**Figures 1** and **2**).

4.3 Alternatives mediated by targeting central OT release

Release from the hypothalamic cell bodies (see **Figure 2**) is indicative of release into the CNS [9]. Such release is also thought to act hormonally to affect surrounding CNS areas [64]. Activation of multiple types of endogenous (ryanodine, cADPr, and opioid) receptors in OT-releasing rat nerve endings can specifically regulate the in vitro release of this neuropeptide [13, 65, 66]. These receptor types play a role in modulating intracellular Ca^{2+} levels, either through a Ca^{2+}-induced Ca^{2+} release (CICR)-dependent mechanism (**Figure 2**) or by regulating Ca^{2+} inflow from the extracellular environment. These receptor types are easily targeted by drugs that can cross the blood–brain barrier [67] and will facilitate novel pharmacological studies with therapeutic relevance for ASD. Furthermore, our collaborators [25, 68, 69] have already shown that the hypothalamic release of OT specifically can be potentiated by other agents that affect $[Ca^{2+}]$, e.g., IP_3 (but not ryanodine). Thus, it should be possible to specifically potentiate/inhibit endogenous levels of central OT in response to physiological stimulations. That is, IP_3R agonists would increase central, local release and electrical activity in OT cell bodies while RyR agonists would increase release from OT terminals both in NH and in central projections.

4.4 Future perspectives

CD38 knockouts (KOs) do not release OT and, consequently, have social behavioral deficits [23]. Similarly, VPA is associated with a high incidence of ASD phenotypes in children [55, 70] and it acts similarly in rats. These KO mice and valproic acid (VPA)-injected rats are extremely useful experimental models for studying potential causes of ASD [47] and exploring new therapeutic approaches.

To target OT vs. AVP release in the relevant areas of the brain [58] for behavioral defects associated with ASD, it will be important to determine if there are differences between central and peripheral OT release using the HNS (**Figure 2**) in a compartmentalized chamber [71] or in fluorescently labeled rats [72]. Next, we have to determine central OT vs. AVP levels in the CD38 KO and VPA autistic animal models. Finally, can known regulators of OT vs. AVP release "correct" OT levels in such models (see **Figure 2**)? These receptor types are easily targeted by drugs that can cross the blood–brain barrier [67] and will facilitate novel studies with therapeutic relevance. Some, or perhaps all, of these agents, will increase OT vs. AVP release, in general, but it should be determined if any (e.g., IP_3, cADPr, mu opioid receptor (MOR) agonists) will specifically increase the central release of OT. Thus, it should make them potential therapeutic agents to test on ASD patients. This approach has been recently validated since a small molecule vasopressin V1a (V1a)-specific antagonist is sufficient to rescue normal behavior in prenatally exposed VPA rats [67]. Furthermore, these findings could eventually lead to the ability to perform clinical trials to assess the efficacy of such agents on ASD patients.

5. Conclusions

We have demonstrated that intracellular signaling pathways in particular CNS areas (see **Figure 2**) can be upregulated. Thus, it should facilitate physiologically stimulated release of OT specifically, rather than nonspecific increases (as with OT injections or sprays). Therefore, facilitating physiological neuronal release of OT but not AVP from the HNS via modulation of IP_3 vs. ryanodine calcium stores (see **Figure 2**) would specifically facilitate central vs. peripheral OT release in ASD patients. Therapeutically, this would ameliorate social behavioral deficits in such patients.

Acknowledgements

We want to thank our many colleagues who helped with this research and the funding sources NIH 1R01AA027808-01A1 (Velázquez) and NIH R01NS29470 (Lemos).

Conflict of interest

The authors declare no conflict of interest.

Appendices and nomenclature

OT	Oxytocin
AVP	Arginine vasopressin
RyR	Ryanodine receptor
IP3R	Inositol trisphosphate receptor
ASD	Autism spectrum disorder

Gq	guanine nucleotide-binding protein (G protein, q polypeptide)
SNARE	soluble N-ethylmaleimide-sensitive factor attachment protein receptors
cADPr	cyclic adenosine diphosphate ribose
CD38	cyclic ADP ribose hydrolase
SON	Supraoptic nucleus
PVN	Paraventricular nucleus
CNS	Central nervous system
NH	Neurohypophysis
HNS	Hypothalamic neurohypophysial system
VGCC	Voltage-gated calcium channel
NSG	Neurosecretory granule
CICR	Calcium-induced calcium release
NHT	Neurohypophysial terminal
ER	Endoplasmic reticulum

Author details

Cristina Velázquez-Marrero[1*] and José R. Lemos[2]

1 UPR-MSC Institute of Neurobiology, San Juan, PR, USA

2 Department of Microbiology and Physiological Systems, University of Massachusetts Chan Medical School, Worcester, MA, USA

*Address all correspondence to: cristina.velazquez1@upr.edu

References

[1] Velazquez-Marrero C et al. Mu-opioid inhibition of Ca^{2+} currents and secretion in isolated terminals of the neurohypophysis occurs via ryanodine-sensitive Ca^{2+} stores. The Journal of Neuroscience. 2014;**34**(10):3733-3742

[2] De Crescenzo V et al. Ca2+ syntillas, miniature Ca2+ release events in terminals of hypothalamic neurons, are increased in frequency by depolarization in the absence of Ca2+ influx. The Journal of Neuroscience. 2004;**24**(5):1226-1235

[3] McNally JM et al. Individual calcium syntillas do not trigger spontaneous exocytosis from nerve terminals of the neurohypophysis. The Journal of Neuroscience. 2009;**29**(45):14120-14126

[4] Kanner L. Autistic disturbances of affective contact. Acta Paedopsychiatrica. 1968;**35**(4):100-136

[5] Yamasue H. Function and structure in social brain regions can link oxytocin-receptor genes with autistic social behavior. Brain and Development. 2013;**35**(2):111-118

[6] Yamasue H et al. Integrative approaches utilizing oxytocin to enhance prosocial behavior: From animal and human social behavior to autistic social dysfunction. The Journal of Neuroscience. 2012;**32**(41):14109-14117

[7] Choleris E et al. Neuroendocrinology of social information processing in rats and mice. Frontiers in Neuroendocrinology. 2009;**30**(4): 442-459

[8] Sala M et al. Pharmacologic rescue of impaired cognitive flexibility, social deficits, increased aggression, and seizure susceptibility in oxytocin receptor null mice: A neurobehavioral model of autism. Biological Psychiatry. 2011;**69**(9):875-882

[9] Veenema AH, Neumann ID. Central vasopressin and oxytocin release: Regulation of complex social behaviours. Progress in Brain Research. 2008;**170**:261-276

[10] Winslow JT, Insel TR. Neuroendocrine basis of social recognition. Current Opinion in Neurobiology. 2004;**14**(2):248-253

[11] Modahl C et al. Plasma oxytocin levels in autistic children. Biological Psychiatry. 1998;**43**(4):270-277

[12] Dayanithi G et al. Segregation of calcium signalling mechanisms in magnocellular neurones and terminals. Cell Calcium. 2012;**51**(3-4):293-299

[13] Lemos JR et al. Modulation/physiology of calcium channel sub-types in neurosecretory terminals. Cell Calcium. 2012;**51**(3-4):284-292

[14] Velazquez-Marrero C et al. Voltage-induced Ca(2+) release by ryanodine receptors causes neuropeptide secretion from nerve terminals. Journal of Neuroendocrinology. 2020;**32**(4):e12840

[15] Fill M, Copello JA. Ryanodine receptor calcium release channels. Physiological Reviews. 2002;**82**(4):893-922

[16] Ortiz-Miranda SI et al. Mu-opioid receptor modulates peptide release from rat neurohypophysial terminals by inhibiting Ca(2+) influx. Journal of Neuroendocrinology. 2003;**15**(9):888-894

[17] De Crescenzo V et al. Dihydropyridine receptors and type 1 ryanodine receptors constitute the molecular machinery for voltage-induced Ca2+ release in nerve terminals. The Journal of Neuroscience. 2006;**26**(29):7565-7574

[18] Wang G, Lemos JR. Tetrandrine blocks a slow, large-conductance, Ca(2+)-activated potassium channel besides inhibiting a non-inactivating Ca2+ current in isolated nerve terminals of the rat neurohypophysis. Pflügers Archiv. 1992;**421**(6):558-565

[19] Galione A, White A. Ca2+ release induced by cyclic ADP-ribose. Trends in Cell Biology. 1994;**4**(12):431-436

[20] Morita K, Kitayama S, Dohi T. Physiological role of cyclic ADP-ribose as a novel endogenous agonist of ryanodine receptor in adrenal chromaffin cells. Nihon Yakurigaku Zasshi. 2002;**120**(1):96P-98P

[21] Sitsapesan R, McGarry SJ, Williams AJ. Cyclic ADP-ribose, the ryanodine receptor and Ca2+ release. Trends in Pharmacological Sciences. 1995;**16**(11):386-391

[22] Higashida H et al. Oxytocin signal and social behaviour: Comparison among adult and infant oxytocin, oxytocin receptor and CD38 gene knockout mice. Journal of Neuroendocrinology. 2010;**22**(5):373-379

[23] Jin D et al. CD38 is critical for social behaviour by regulating oxytocin secretion. Nature. 2007;**446**(7131):41-45

[24] Lee HC, Aarhus R. Fluorescent analogs of NAADP with calcium mobilizing activity. Biochimica et Biophysica Acta. 1998;**1425**(1):263-271

[25] Ludwig M et al. Intracellular calcium stores regulate activity-dependent neuropeptide release from dendrites. Nature. 2002;**418**(6893):85-89

[26] Tobin VA et al. Thapsigargin-induced mobilization of dendritic dense-cored vesicles in rat supraoptic neurons. The European Journal of Neuroscience. 2004;**19**(10):2909-2912

[27] Ludwig M et al. Regulation of activity-dependent dendritic vasopressin release from rat supraoptic neurones. The Journal of Physiology. 2005;**564**(Pt 2):515-522

[28] Dayanithi G, Widmer H, Richard P. Vasopressin-induced intracellular Ca2+ increase in isolated rat supraoptic cells. The Journal of Physiology. 1996;**490**(Pt 3):713-727

[29] McNally JM et al. Functional ryanodine receptors in the membranes of neurohypophysial secretory granules. The Journal of General Physiology. 2014;**143**(6):693-702

[30] Debiec J. Peptides of love and fear: Vasopressin and oxytocin modulate the integration of information in the amygdala. BioEssays. 2005;**27**(9):869-873

[31] Li YJ, Du WJ, Liu R, Zan GY, Ye BL, Li Q, et al. Paraventricular nucleus-central amygdala oxytocinergic projection modulates pain-related anxiety-like behaviors in mice. CNS Neuroscience & Therapeutics. 29 May 2023. DOI: 10.1111/cns.14282. Epub ahead of print. PMID: 37248645

[32] Leng G, Brown D. The origins and significance of pulsatility in hormone secretion from the pituitary. Journal of Neuroendocrinology. 1997;**9**(7):493-513

[33] Richard P et al. Rhythmic activities of hypothalamic magnocellular neurons:

Autocontrol mechanisms. Biology of the Cell. 1997;**89**(9):555-560

[34] Rossoni E et al. Emergent synchronous bursting of oxytocin neuronal network. PLoS Computational Biology. 2008;**4**(7):e1000123

[35] Dyball RE, Leng G. Regulation of the milk ejection reflex in the rat. The Journal of Physiology. 1986;**380**:239-256

[36] Russell JA, Leng G, Douglas AJ. The magnocellular oxytocin system, the fount of maternity: Adaptations in pregnancy. Frontiers in Neuroendocrinology. 2003;**24**(1):27-61

[37] Summerlee AJ. Hypothalamic neurone activity: Hormone release and behaviour in freely moving rats. Quarterly Journal of Experimental Physiology. 1983;**68**(3):505-515

[38] Armstrong WE et al. Electrophysiological properties of identified oxytocin and vasopressin neurones. Journal of Neuroendocrinology. 2019;**31**(3): e12666

[39] Ortiz-Miranda S et al. Micro-opioid receptor preferentially inhibits oxytocin release from neurohypophysial terminals by blocking R-type Ca2+ channels. Journal of Neuroendocrinology. 2005;**17**(9):583-590

[40] Wang G et al. An R-type Ca(2+) current in neurohypophysial terminals preferentially regulates oxytocin secretion. The Journal of Neuroscience. 1999;**19**(21):9235-9241

[41] Kirchner MK et al. Phosphatidylinositol 4,5-bisphosphate (PIP(2)) modulates afterhyperpolarizations in oxytocin neurons of the supraoptic nucleus. The Journal of Physiology. 2017;**595**(14):4927-4946

[42] Gamper N et al. Phosphatidylinositol [correction] 4,5-bisphosphate signals underlie receptor-specific Gq/11-mediated modulation of N-type Ca2+ channels. The Journal of Neuroscience. 2004;**24**(48):10980-10992

[43] Chevaleyre V et al. Developmental regulation of a local positive autocontrol of supraoptic neurons. The Journal of Neuroscience. 2000;**20**(15):5813-5819

[44] Da Costa AP et al. The role of oxytocin release in the paraventricular nucleus in the control of maternal behaviour in the sheep. Journal of Neuroendocrinology. 1996;**8**(3):163-177

[45] Zhong M, Yang M, Sanborn BM. Extracellular signal-regulated kinase 1/2 activation by myometrial oxytocin receptor involves Galpha(q) Gbetagamma and epidermal growth factor receptor tyrosine kinase activation. Endocrinology. 2003;**144**(7):2947-2956

[46] Parnas I, Parnas H. Calcium is essential but insufficient for neurotransmitter release: The calcium-voltage hypothesis. Journal of Physiology, Paris. 1986;**81**(4):289-305

[47] Bambini-Junior V et al. Animal model of autism induced by prenatal exposure to valproate: Behavioral changes and liver parameters. Brain Research. 2011;**1408**:8-16

[48] Gadia CA, Tuchman R, Rotta NT. Autism and pervasive developmental disorders. Jornal de Pediatria. 2004;**80**(2 Suppl):S83-S94

[49] Green JJ, Hollander E. Autism and oxytocin: New developments in translational approaches to therapeutics. Neurotherapeutics. 2010;**7**(3):250-257

[50] Kogan MD et al. Prevalence of parent-reported diagnosis of

autism spectrum disorder among children in the US, 2007. Pediatrics. 2009;**124**(5):1395-1403

[51] Leithead AB, Tasker JG, Harony-Nicolas H. The interplay between glutamatergic circuits and oxytocin neurons in the hypothalamus and its relevance to neurodevelopmental disorders. Journal of Neuroendocrinology. 2021;**33**(12):e13061

[52] Rapin I, Tuchman RF. What is new in autism? Current Opinion in Neurology. 2008;**21**(2):143-149

[53] Muhle R, Trentacoste SV, Rapin I. The genetics of autism. Pediatrics. 2004;**113**(5):e472-e486

[54] Bello SC. Autism and environmental influences: Review and commentary. Reviews on Environmental Health. 2007;**22**(2):139-156

[55] Christianson AL, Chesler N, Kromberg JG. Fetal valproate syndrome: Clinical and neuro-developmental features in two sibling pairs. Developmental Medicine and Child Neurology. 1994;**36**(4):361-369

[56] Fombonne E. Epidemiology of pervasive developmental disorders. Pediatric Research. 2009;**65**(6):591-598

[57] Rodier PM et al. Linking etiologies in humans and animal models: Studies of autism. Reproductive Toxicology. 1997;**11**(2-3):417-422

[58] Smeltzer MD et al. Dopamine, oxytocin, and vasopressin receptor binding in the medial prefrontal cortex of monogamous and promiscuous voles. Neuroscience Letters. 2006;**394**(2):146-151

[59] Guastella AJ et al. Does oxytocin influence the early detection of angry and happy faces? Psychoneuroendocrinology. 2009;**34**(2):220-225

[60] Lee HJ et al. A conditional knockout mouse line of the oxytocin receptor. Endocrinology. 2008;**149**(7):3256-3263

[61] Park J et al. Evidence that genetic variation in the oxytocin receptor (OXTR) gene influences social cognition in ADHD. Progress in Neuro-Psychopharmacology & Biological Psychiatry. 2010;**34**(4):697-702

[62] Haas BW, Smith AK. Oxytocin, vasopressin, and Williams syndrome: Epigenetic effects on abnormal social behavior. Frontiers in Genetics. 2015;**6**:28

[63] Henrichsen CN et al. Using transcription modules to identify expression clusters perturbed in Williams-Beuren syndrome. PLoS Computational Biology. 2011;**7**(1):e1001054

[64] Leng G, Ludwig M. Neurotransmitters and peptides: Whispered secrets and public announcements. The Journal of Physiology. 2008;**586**(23):5625-5632

[65] Custer EE et al. Effects of calcium and sodium on ATP-induced vasopressin release from rat isolated neurohypophysial terminals. Journal of Neuroendocrinology. 2018;**30**:1-8

[66] Ortiz-Miranda S, McNally J, Custer E, Salzberg BM, Woodbury D, Lemos JR. Functional ryanodine receptors in the membranes of neurohypophysial secretory granules. Society for Neuroscience - Abstracts. 2011;**36**(446.24)

[67] Umbricht D et al. A single dose, randomized, controlled proof-of-mechanism study of a novel vasopressin 1a receptor antagonist (RG7713) in high-functioning adults with autism Spectrum disorder. Neuropsychopharmacology. 2017;**42**(9):1914-1923

[68] Sabatier N, Richard P, Dayanithi G. Activation of multiple intracellular transduction signals by vasopressin in vasopressin-sensitive neurones of the rat supraoptic nucleus. The Journal of Physiology. 1998;**513**(Pt 3):699-710

[69] Viero C, Dayanithi G. Neurosteroids are excitatory in supraoptic neurons but inhibitory in the peripheral nervous system: It is all about oxytocin and progesterone receptors. Progress in Brain Research. 2008;**170**:177-192

[70] Bromley RL et al. Early cognitive development in children born to women with epilepsy: A prospective report. Epilepsia. 2010;**51**(10):2058-2065

[71] Rossi NF. Regulation of vasopressin secretion by ETA and ETB receptors in compartmentalized rat hypothalamo-neurohypophysial explants. American Journal of Physiology. Endocrinology and Metabolism. 2004;**286**(4):E535-E541

[72] Ueta Y, Dayanithi G, Fujihara H. Hypothalamic vasopressin response to stress and various physiological stimuli: Visualization in transgenic animal models. Hormones and Behavior. 2011;**59**(2):221-226

Chapter 2

The Love Hormone and Seizure Control: A Review of Oxytocin's Impact on Epilepsy Management

Lufuno Makhado and Thendo Gertie Makhado

Abstract

Epilepsy is a neurological disorder characterised by recurrent seizures, which can significantly impact patient's quality of life. While current management strategies for epilepsy, such as antiepileptic drugs and surgery, are effective for many patients, there is a need for novel therapies that can provide better seizure control and improve patients' outcomes. Oxytocin, a neuropeptide known for its role in social bonding and trust, has emerged as a promising therapy for epilepsy. Preclinical studies have shown that oxytocin can reduce seizure activity and improve seizure outcomes in animal models of epilepsy. In contrast, clinical studies have suggested that oxytocin may reduce seizure frequency and severity in some epilepsy patients. This chapter reviews the current knowledge of oxytocin and epilepsy, including the potential mechanisms of oxytocin's antiepileptic effects, the limitations and challenges of clinical studies, and future research directions and implications. The chapter also discusses the broader impact of oxytocin research on understanding social behaviour and neurological disorders. Overall, the chapter highlights the potential of oxytocin as a novel therapy for epilepsy management and underscores the need for further research.

Keywords: oxytocin, seizure control, epilepsy management, improved patient outcomes, epilepsy

1. Introduction

Oxytocin, often called the "love hormone," has garnered considerable attention for its profound effects on social behaviours and emotional bonding [1–5]. It fosters trust, empathy, and social connection [3, 4, 6]. However, recent research has uncovered additional dimensions of oxytocin's influence, particularly in neurological disorders [7–9]. One such disorder is epilepsy, a condition characterised by recurrent seizures resulting from abnormal electrical activity in the brain. Understanding the potential role of oxytocin in modulating seizure control and its broader implications for individuals with epilepsy opens new avenues for therapeutic interventions.

Epilepsy affects millions of people worldwide, and while traditional antiepileptic drugs have successfully managed seizures for many individuals, a significant portion of patients continue to experience seizures despite optimal medical management [10, 11]. This treatment-resistant epilepsy poses challenges and underscores the need to explore alternative therapeutic approaches. In this context, oxytocin has emerged as a promising candidate due to its multifaceted effects on the brain and potential to influence seizure activity.

Beyond oxytocin's prominent role in facilitating childbirth and lactation, oxytocin is now recognised as a critical neuromodulator involved in various physiological and cognitive processes. Extensive research has revealed its effects on stress regulation, emotional processing, and social cognition [12]. These effects are mediated through oxytocin receptors distributed in key brain regions implicated in epilepsy, including the hippocampus and amygdala [13]. Consequently, researchers have begun investigating oxytocin's potential impact on epilepsy, exploring its ability to modulate seizure activity and enhance seizure control outcomes.

By exploring the relationship between oxytocin and epilepsy, this chapter aims to provide a comprehensive review of the current understanding of oxytocin's impact on epilepsy management. We will delve into the underlying mechanisms through which oxytocin may modulate seizures, including its effects on neuronal excitability, synaptic transmission, and network synchronisation. Additionally, we will examine the potential of oxytocin as an adjunctive treatment for epilepsy, considering its ability to reduce seizure frequency, improve seizure control outcomes, and enhance the response to conventional antiepileptic drugs. Understanding the potential therapeutic implications of oxytocin in epilepsy management can pave the way for novel treatment approaches and personalised interventions, ultimately improving the quality of life for individuals with epilepsy.

2. Background

Oxytocin, a peptide hormone synthesised in the hypothalamus and released by the pituitary gland, has long been recognised for its crucial role in childbirth and lactation, facilitating maternal-infant bonding and promoting nurturing behaviours [14–16]. However, oxytocin's influence extends far beyond the reproductive realm to physiological, neurological, and cognitive processes. In recent years, researchers have discovered that oxytocin acts as a versatile neuromodulator, exerting effects on various physiological and cognitive processes [17–20].

Studies have demonstrated oxytocin's involvement in stress regulation, anxiety modulation, and emotional processing. Oxytocin enhances social bonding among human beings by promoting trust, empathy, and prosocial behaviours [21–25]. It influences social cognition, including the perception of facial expressions, emotions, and social cues [26–34]. This reveals that oxytocin regulates social reward and reinforcement, impacting social motivation and attachment.

The involvement of oxytocin in epilepsy, a neurological condition characterised by recurring seizures, has attracted the curiosity of academics and physicians. Epilepsy arises from abnormal electrical activity in the brain, resulting in the production and spread of seizures. While traditional antiepileptic medications have successfully treated seizures for many people, many patients suffer from treatment-resistant epilepsy, in which seizures continue despite appropriate medical therapy [35, 36]. This treatment gap necessitates exploring alternative therapeutic approaches, and oxytocin has emerged as a potential candidate.

The ability of oxytocin to modulate seizure control originates from its impact on the central nervous system (CNS) [37, 38]. Oxytocin receptors, including the hippocampus and amygdala, are present throughout the brain and have been linked to both pro- and anticonvulsive characteristics, which appear to be dose- or time-dependent [39, 40]. Preclinical studies utilising animal models of epilepsy have shown that oxytocin administration can reduce seizure frequency and severity. Oxytocin acts as an anticonvulsant via various methods, including the modulation of neurotransmitter systems such as gamma-aminobutyric acid (GABA) and glutamate and regulation of ion channels involved in neural excitability [41, 42].

Understanding the potential role of oxytocin in epilepsy care has important implications for improving seizure control results and improving the quality of life for people living with epilepsy. Furthermore, understanding the complicated connection between oxytocin, social behaviour, and epilepsy may shed light on the complex interplay between neurological illnesses and social deficits. This chapter opens the door for possible oxytocin therapeutic applications in seizure control and ameliorating social deficiencies commonly reported in epileptic patients. This chapter attempts to completely comprehend oxytocin's impact on epilepsy management by evaluating the available literature and highlighting its potential as a helpful therapeutic tool in epilepsy research and clinical practice.

2.1 Purpose of the chapter

The primary objective of this chapter is to provide a comprehensive understanding of the impact of oxytocin on epilepsy management. Reviewing the existing literature, we aim to elucidate the intricate relationship between oxytocin and seizure control, exploring oxytocin's potential as a therapeutic tool. The chapter delves into the mechanisms by which oxytocin modulates seizures, its effects on seizure generation and propagation, and its potential as an adjunctive treatment for epilepsy management.

3. Oxytocin: the love hormone

This section described oxytocin and its physiological functions, particularly its role in social bonding and trust, as well as the mechanisms of oxytocin release and its effects on brain activity.

3.1 Oxytocin and its physiological functions, particularly its role in social bonding and trust

Oxytocin acts as a neurotransmitter and neuromodulator, influencing various brain regions involved in social behaviours and emotional processing. It binds to specific receptors in the brain, particularly in areas such as the amygdala, hippocampus, and prefrontal cortex, which are critical for regulating emotions and social interactions.

The role of oxytocin in social bonding and trust has been extensively studied. Oxytocin promotes bonding between individuals, particularly in close relationships such as romantic partners, parents and children, and friends. It enhances attachment and fosters intimacy and connection [43, 44]. Studies have shown that oxytocin increases feelings of trust and cooperation in social interactions, leading to more positive social behaviours [21–25, 45, 46]. It promotes empathy and facilitates the ability to understand and share the emotions of others, which is vital for building and maintaining social relationships [24, 47, 48].

Furthermore, oxytocin has been implicated in the regulation of stress and anxiety. It acts as a natural stress buffer, helping reduce stress responses and promote calmness and relaxation in social situations [44, 49–51]. Oxytocin's anti-anxiety effects contribute to its role in social bonding and trust, as it helps to create a sense of safety and security in interpersonal interactions.

The effects of oxytocin on social bonding and trust are not limited to human interactions but extend to other species. Oxytocin has been shown to play a crucial role in maternal bonding in mammals, promoting the maternal-infant bond and facilitating nurturing behaviours [52–54]. It also influences social behaviours in non-human animals, fostering affiliative behaviours, pair bonding, and cooperative interactions within social groups.

Thus, oxytocin is a neuropeptide hormone that plays a significant role in social behaviours and emotional bonding. It promotes social bonding, trust, and empathy, facilitating the formation and maintenance of close relationships. Oxytocin enhances prosocial behaviours, reduces social threat responses, and regulates stress. Understanding the physiological functions of oxytocin and its role in social bonding and trust provides valuable insights into the complex mechanisms underlying human and animal social interactions.

3.2 The mechanisms of oxytocin release and its effects on brain activity

A complex interplay of physiological and psychological factors regulates the release of oxytocin. In response to various stimuli, including positive social interactions, touch, and emotional cues, oxytocin is released into the bloodstream and acts as a neurotransmitter in the brain. Oxytocin release can be triggered by childbirth, breastfeeding, and intimate physical contact, reinforcing individual bonding [55–58].

Oxytocin affects brain activity by interacting with specific oxytocin receptors distributed throughout the central nervous system, including regions involved in social behaviours and emotional processing. Multiple pathways mediate the effects of oxytocin on brain activity. Oxytocin receptors are densely expressed in brain regions such as the amygdala, hippocampus, and prefrontal cortex, which are involved in emotional regulation, social cognition, and memory formation [59–62].

Upon binding to oxytocin receptors, oxytocin triggers a cascade of intracellular signalling pathways that modulate neuronal excitability, synaptic transmission, and network connectivity [63–66]. Oxytocin enhances the release of inhibitory neurotransmitters, such as GABA, which dampen neural activity and reduce excitability [41, 42]. This inhibitory effect of oxytocin helps regulate emotional responses and may contribute to its anxiolytic properties.

Furthermore, oxytocin influences the processing of social stimuli and the interpretation of social cues. It enhances social information's salience and reward value, making social stimuli more rewarding and reinforcing social bonding [21–25, 67–72]. Oxytocin also modulates the activity of brain regions involved in empathy and social cognition, promoting the understanding of others' emotions and intentions [73–77].

In summary, oxytocin is a neuropeptide hormone that plays a vital role in social bonding and trust. It is released in response to positive social interactions and acts on specific oxytocin receptors in the brain. Oxytocin modulates neural activity, enhances inhibitory neurotransmission, and influences the processing of social stimuli, ultimately promoting social bonding, trust, and prosocial behaviours. Understanding the mechanisms of oxytocin release and its effects on brain activity provides valuable insights into the neurobiology of social behaviour and its potential therapeutic applications in various domains, including epilepsy management and social deficits in neurological disorders.

4. Epilepsy: overview and current management strategies

Epilepsy, a chronic neurological disorder characterised by unprovoked seizures, affects approximately 1% of the global population [78]. Seizures can take many forms, from minor changes in awareness or sensation to more severe convulsions and loss of consciousness [79, 80]. Recognition of epileptic symptoms is crucial for early detection and appropriate care. The symptoms can vary depending on the brain region abnormal electrical activity affects [81–83]. Some individuals may experience brief episodes of altered consciousness, such as staring or repetitive movements, while others may have more severe seizures involving convulsions and loss of muscle control [84–88].

Epilepsy is caused by various factors that differ from person to person. Epilepsy is frequently associated with structural abnormalities in the brain, such as deformities, injuries, tumours, strokes, or infections [84–88]. In some cases, genetic or hereditary factors contribute to the development of epilepsy, with specific gene mutations or inherited disorders increasing seizure susceptibility [79, 89–91]. Metabolic abnormalities, including electrolyte imbalances or glucose metabolism disorders, can also play a role in epilepsy [92–94]. Nonetheless, many cases of epilepsy have no known causes and are classified as idiopathic or cryptogenic epilepsy [79, 95, 96].

The primary goal of epilepsy management is to control seizures, minimise their impact on daily life, and enhance the overall quality of life for individuals with the condition. Antiepileptic drugs (AEDs) are the cornerstone of current management strategies, aiming to regulate brain electrical activity and reduce the frequency and severity of seizures [97–100]. AEDs are chosen based on the type of seizure, epilepsy syndrome, age, and individual patient characteristics [79, 101, 102]. Lifestyle modifications, including regular sleep patterns, stress management, adherence to medication regimens, and routine medical monitoring, may complement medication-based management [103–107].

Despite the efficacy of current management strategies, more research and innovation are required to improve treatment outcomes and quality of life for people living with epilepsy. For individuals who do not respond adequately to medication, researchers are investigating novel therapeutic approaches, such as neuromodulation techniques (e.g., vagus nerve stimulation and deep brain stimulation), dietary therapies (e.g., ketogenic diet), and surgical interventions [79]. Neuromodulation techniques, such as vagus nerve stimulation, have shown promising results in reducing seizure frequency and improving seizure control [79, 108–112]. Some patients, particularly those with refractory epilepsy, have shown efficacy in dietary therapies such as the ketogenic diet [113–117]. Individuals with focal epilepsy who do not respond to medication may be candidates for surgical interventions such as resective surgery or responsive neurostimulation [118–123].

Diagnostic procedures are being refined as well to improve accuracy and efficiency. Magnetic resonance imaging (MRI) and positron emission tomography (PET) are advanced neuroimaging techniques that provide valuable insights into structural and functional abnormalities in the brain associated with Refs. [124–127]. Genetic profiling is increasingly utilised to identify specific gene mutations or genetic variants contributing to epilepsy susceptibility, allowing for more personalised treatment approaches [128–132].

Epilepsy is a complicated neurological disorder characterised by recurrent and unprovoked seizures caused by abnormal electrical activity in the brain. Understanding the symptoms, causes, and current management techniques is critical for timely diagnosis, effective treatment, and improving the quality of life for people living with epilepsy. Antiepileptic drugs are the foundation of therapy. However, ongoing research is focused on developing novel therapeutic approaches to improve seizure control and management, such as neuromodulation techniques and dietary therapies. Furthermore, advances in diagnostic procedures, such as advanced neuroimaging techniques and genetic profiling, hold promise for personalised epilepsy treatment strategies. These developments aim to improve diagnostic accuracy, identify genetic factors contributing to epilepsy susceptibility, and facilitate personalised treatment plans. To control seizures, reduce their impact on daily life, and improve the overall well-being of people living with epilepsy, comprehensive management strategies are required. Healthcare professionals can improve treatment outcomes and patients' quality of life through the combination of pharmacological interventions, lifestyle changes, and emerging therapeutic options. Continuous research is required to enhance and strengthen epilepsy management and further understand this complex neurological condition.

Furthermore, future improvements in diagnostic procedures, such as improved neuroimaging techniques and genetic profiling, embrace the promise of personalised epilepsy treatment strategies. These developments aim to improve diagnostic accuracy, identify genetic factors contributing to epilepsy susceptibility, and facilitate personalised treatment plans. To control seizures, reduce their impact on daily life, and improve the overall well-being of people living with epilepsy, comprehensive management strategies are required. Healthcare professionals can improve treatment outcomes and patients' quality of life by combining pharmacological interventions, lifestyle changes, and emerging therapeutic options. Continuous research is required to enhance epilepsy management and further our understanding of this complex neurological condition.

4.1 Epilepsy symptoms and causes

Epilepsy is a neurological condition characterised by two or more spontaneous seizures [133, 134]. Seizures are brief interruptions in normal brain function that result from abnormal electrical activity [135, 136]. Seizures can cause various symptoms, ranging from transient changes in awareness or sensation to convulsions and loss of consciousness, depending on the specific region of the brain affected [85, 137, 138].

The causes of epilepsy can be categorised into structural and non-structural factors. Structural causes are associated with identifiable brain abnormalities or lesions [139, 140]. These abnormalities include cortical malformations resulting from irregular migration of brain cells during foetal development and brain damage caused by traumatic brain injuries [139, 140]. Brain tumours, strokes, and central nervous system infections, such as meningitis or encephalitis, can potentially lead to epilepsy [139, 141, 142].

Alternatively, non-structural causes of epilepsy do not involve apparent structural abnormalities within the brain but are often linked to underlying functional or genetic factors [79, 143–145]. Genetic mutations or inherited diseases can predispose individuals to seizures, highlighting the role of hereditary factors in epilepsy [79, 146, 147].

Furthermore, conditions, such as autism spectrum disorder and intellectual difficulties, are associated with an increased risk of seizures [79, 146, 147]. Metabolic disorders can also trigger seizures, including electrolyte imbalances and glucose metabolism abnormalities [79, 92, 94, 148].

Elaborately, epilepsy is characterised by two or more spontaneous seizures resulting from abnormal electrical activity in the brain. The causes of epilepsy can be divided into structural factors, including identifiable brain abnormalities or lesions, and non-structural factors, associated with underlying functional or genetic factors. Understanding the various causes of epilepsy is critical for precise diagnosis and effective management of the condition.

It is necessary to highlight that the precise cause of many cases of epilepsy is unknown. Idiopathic or cryptogenic epilepsy is the medical term for this condition. Despite developments in diagnostic tools and understanding of the illness, the underlying cause of epilepsy remains a mystery in many patients and the general public [11]. Ongoing research seeks to identify the genetic and molecular factors that contribute to the development of epilepsy to improve diagnosis, treatment, and management strategies. Ultimately, epilepsy is a complex neurological illness marked by recurring seizures. Epilepsy can be caused by structural abnormalities in the brain or by non-structural factors such as functional or hereditary variables. While breakthroughs in understanding have allowed identifiable causes to be identified in some instances, the precise aetiology of epilepsy remains. Further studies are needed to elucidate the underlying mechanisms and create specific treatments for diverse subtypes of epilepsy, eventually enhancing the management and quality of life for people suffering from this disorder.

4.2 Current management strategies for epilepsy

The treatment of epilepsy is multidimensional, anticipating achieving optimal seizure control, minimising treatment adverse effects, and enhancing the overall quality of life for those living with epilepsy. The foremost method of managing epilepsy is using antiepileptic drugs (AEDs). These drugs work in the brain by modifying neuronal excitability and suppressing aberrant electrical activity. AEDs are chosen depending on various parameters, including seizure type, epilepsy syndrome, and particular patient features. The goal is to determine the best effective AED for each patient, delivering appropriate seizure control while minimising side effects [149–151]. AEDs are the foundation of epilepsy treatment, and various alternatives have diverse modes of action and efficacy profiles. Seizure type, epileptic syndrome, comorbidities, potential drug interactions, and patient preferences all influence AED selection. The ultimate goal is to determine the best appropriate AED or combination of AEDs to achieve optimal seizure control while maintaining the patient's quality of life.

Despite the large variety of AEDs available, roughly 30% of people living with epilepsy do not acquire appropriate seizure control with these drugs [11]. If AEDs fail to manage seizures, alternative treatment options may be tried. Surgical intervention is one such method. Surgical alternatives include excision of the epileptic focus, which involves surgically removing the same brain region responsible for seizures, or the implantation of neurostimulation devices such as vagus nerve stimulation or responsive neurostimulation. When seizures originate from a well-defined brain region that does not impair essential brain functions, surgical procedures are often considered [152]. These surgical methods seek to eliminate or modify the epileptogenic zone, lowering seizure activity and enhancing seizure management overall [153].

While AEDs and surgical treatments have helped many people living with epilepsy, it is crucial to recognise that these options have limitations. Adverse effects of AEDs include cognitive impairment, emotional disturbances, and probable teratogenicity in women of reproductive age. Furthermore, surgical interventions are inappropriate for many patients and may pose risks due to the invasive nature of the procedures and the possibility of consequences.

To summarise, managing epilepsy entails a multifaceted strategy for obtaining optimal seizure control and increasing the overall quality of life for people living with epilepsy. The primary treatment method is antiepileptic medicines; AEDs are chosen based on the seizure type, epilepsy syndrome, and specific patient features. Surgical treatments may be explored if AEDs fail to offer adequate seizure control. However, even current therapy modalities have limits, such as the potential adverse effects of AEDs and the invasiveness and appropriateness of surgical approaches. Continued research and developments in epilepsy management are required to address these limitations and give individuals with epilepsy more effective and personalised treatment alternatives.

Limitations of Current Management Strategies and the Need for Novel Therapies:

While antiepileptic drugs (AEDs) and surgical treatments have successfully treated seizures for many people living with epilepsy, current care techniques have substantial limits. Around 30–40% of patients do not achieve seizure control using existing AEDs [11, 84]. This underscores the need for alternate therapeutic alternatives that can give these people better seizure control. Furthermore, AEDs can cause cognitive deficits, mood abnormalities, and teratogenicity in women of childbearing age, negatively influencing the quality of life of epilepsy patients [154, 155].

For some patients with well-defined seizure origins that do not involve essential brain functions, surgical procedures such as excision of the epileptic focus are successful [156, 157]. However, these procedures are not appropriate for everyone due to the intrusive nature of the surgeries and the possible hazards associated. Furthermore, in some circumstances, pinpointing the particular epileptic centre may be difficult, making surgical procedures less possible or practicable [156, 158–160]. Furthermore, the fundamental mechanisms and specific causes of epilepsy are yet unknown. This knowledge gap impedes the development of targeted medicines tailored to particular patients' underlying diseases. Because epilepsy is a complex illness with multiple aetiologies, individualised therapy techniques that target the underlying causes are critical for enhancing treatment outcomes.

Comorbidities linked with epilepsy, such as cognitive deficits, mental problems, and social limits, provide considerable obstacles for patients in addition to seizure control. Current management options frequently focus solely on seizure control, leaving out more significant aspects of patient well-being and quality of life. There is an urgent need for medicines that target seizure control while also addressing the cognitive, psychological, and social elements of epilepsy, hence improving overall patient outcomes and everyday functioning. As a result, novel treatment techniques for managing epilepsy are desperately needed. Non-pharmacological interventions, such as neuromodulation techniques (e.g., transcranial magnetic stimulation and vagus nerve stimulation) and dietary therapies (e.g., ketogenic diet), are being investigated, as are precision medicine approaches to match treatments with specific epilepsy subtypes and genetic profiles. We can considerably improve the management of epilepsy and improve the quality of life for people living with it by expanding our understanding of the underlying mechanisms and discovering novel treatment options.

5. Oxytocin and epilepsy: preclinical studies

Preclinical animal model research has shed light on oxytocin's potential antiepileptic benefits and underlying processes. These research studies have shed light on the effects of oxytocin on seizure activity and its potential as a treatment agent for epilepsy.

Oxytocin has been shown to have antiepileptic properties in preclinical forms of epilepsy. For example, oxytocin treatment reduced seizure frequency and intensity in mouse models of temporal lobe epilepsy [161]. Similarly, oxytocin administration inhibited seizure progression and lowered seizure severity in kindling models, which include the progressive development of seizures [162–164]. These data show that oxytocin may have antiepileptic properties. The mechanisms that underpin oxytocin's antiepileptic actions are complex. Modulation of GABAergic transmission is one hypothesised method. GABA is the brain's principal inhibitory neurotransmitter, and an imbalance in inhibitory and excitatory neurotransmission contributes to seizure production. Oxytocin has been found to improve GABAergic inhibitory signalling by increasing GABA release and enhancing GABAergic interneuron activity [165]. This regulation of GABAergic transmission can attenuate neuronal excitability and minimise seizure production.

Inhibiting proinflammatory cytokines is another way by which oxytocin exerts its antiepileptic effects. Proinflammatory cytokines enhance neuronal hyperexcitability and seizure production and have been linked to the pathophysiology of epilepsy. In animal models of epilepsy, oxytocin has been demonstrated to reduce the release of proinflammatory cytokines such as interleukin-1 (IL-1) and tumour necrosis factor-alpha (TNF-α) [166]. Oxytocin may help to decrease seizure activity by lowering neuroinflammation. Furthermore, oxytocin may influence other neurotransmitter systems involved in epilepsy. It has been demonstrated to interact with glutamatergic and dopaminergic signalling, which is essential in excitatory neurotransmission and seizure production [167]. The actions of oxytocin on these systems may contribute to its antiepileptic characteristics.

Thus, preclinical investigations in animal models have shown that oxytocin may have antiepileptic properties. The use of oxytocin has been demonstrated to reduce seizure frequency and severity. Modulation of GABAergic transmission, inhibition of proinflammatory cytokines, and potential interactions with other neurotransmitter systems are the mechanisms behind these effects. These findings support further research into oxytocin as a therapeutic intervention for epilepsy and show its potential as a unique approach to seizure management.

6. Oxytocin and epilepsy: clinical studies

Clinical investigations on the effects of oxytocin on seizure frequency, seizure severity, and quality of life in epilepsy patients have provided valuable insights into oxytocin's potential therapeutic applications in epilepsy management. These studies, however, have limitations and constraints that must be recognised. Several clinical trials have looked into the effect of oxytocin on seizure control in epileptic patients. McCormick [168], for example, studied the effects of intranasal oxytocin delivery in patients with refractory focal epilepsy. When compared to the placebo group, the oxytocin-treated group had a significant reduction in seizure frequency. Similarly, Wang et al. [169] reported that intranasal oxytocin reduced seizure intensity in patients with refractory epilepsy.

Aside from controlling seizures, clinical studies have examined the impact of oxytocin on the quality of life of people living with epilepsy. The effects of intranasal oxytocin on social cognition and quality of life in individuals with temporal lobe epilepsy were reported [170]. The results indicated that social cognition and overall quality of life improved after administering oxytocin. Other studies have investigated the impact of oxytocin on epilepsy and quality of life [171, 172]. For example, a study published in Nature in 2023 found that medial prefrontal cortex oxytocin can mitigate epilepsy and cognitive impairments induced by traumatic brain injury by reducing neuroinflammation in mice [173].

Despite these encouraging findings, clinical trials on oxytocin in epilepsy have limitations and obstacles. One critical problem is that many of these studies have relatively small sample sizes, which may restrict the generalisability of the conclusions. Furthermore, there is a lack of established dosing methods for oxytocin administration, with dose, frequency, and duration of treatment varying among the studies [174]. Because of this heterogeneity, comparing and drawing definitive conclusions from the results is difficult.

Furthermore, the diversity of epileptic patients regarding seizure forms, epilepsy syndromes, and underlying causes hampers the interpretation of findings. Patient variables, such as age, gender, and comorbidities, may alter oxytocin treatment response. These variables emphasise the need for well-designed, multicentre, randomised controlled studies with higher sample sizes and standardised methodologies to determine oxytocin's efficacy, safety, and appropriate dose regimens in epilepsy therapy.

On the other hand, the long-term and potentially harmful consequences of persistent oxytocin delivery in epilepsy patients must be studied further. Oxytocin is a complicated hormone with systemic effects in addition to its involvement in social behaviours, and a careful monitoring is required to assess its safety profile in the context of epilepsy treatment.

It can be concluded that the clinical research on oxytocin in epilepsy has shown encouraging results in terms of seizure control and quality of life. These studies, however, have limitations due to small sample sizes, a lack of standardised dosing regimens, and the variability of epilepsy patients. Addressing these limitations and doing additional studies using rigorous methodology will be critical in identifying oxytocin's therapeutic potential in epilepsy care and determining its appropriate clinical application.

7. Conclusion

Epilepsy is a complex neurological disorder characterised by recurrent and unprovoked seizures caused by abnormal electrical activity in the brain. Understanding the symptoms, causes, and current management techniques is critical for timely diagnosis, effective treatment, and improving epilepsy patients' quality of life. The primary goal of epilepsy management is to control seizures, reduce their impact on daily life, and improve overall well-being. Antiepileptic drugs (AEDs) are currently used to regulate brain electrical activity and reduce the frequency and severity of seizures. Lifestyle changes, such as regular sleep patterns, stress management, adherence to medication regimens, and regular medical monitoring, can supplement medication-based management.

Despite the effectiveness of current management strategies, some limitations and challenges must be addressed. Approximately 30% of patients do not achieve adequate seizure control with AEDs, highlighting the need for alternative therapeutic options. Surgical interventions are considered for individuals who do not respond to medication, but these procedures are inappropriate for all patients and carry risks. Adverse effects of AEDs, such as cognitive impairment and emotional disturbances, can significantly impact the quality of life of epilepsy patients. Furthermore, current management approaches frequently focus solely on seizure control, ignoring critical aspects of patient well-being and quality of life.

Continuous research and innovation are required to overcome these limitations and provide better care for people living with epilepsy. Neuromodulation techniques (e.g., vagus nerve stimulation and deep brain stimulation) and dietary therapies (e.g., ketogenic diet) are being investigated as novel therapeutic approaches, with promising results in reducing seizure frequency and improving seizure control. These new approaches may provide new options for patients who do not respond well to medication.

Furthermore, advances in diagnostic procedures, such as advanced neuroimaging techniques and genetic profiling, hold promise for personalised epilepsy treatment strategies. Healthcare professionals can develop personalised treatment plans that target the underlying causes of epilepsy by improving diagnostic accuracy and identifying specific genetic factors contributing to epilepsy susceptibility. This personalised approach can potentially improve treatment outcomes and the overall well-being of people living with epilepsy.

Comprehensive management strategies should consider the broader aspects of patient well-being in addition to novel therapeutic approaches and personalised medicine. This includes addressing epilepsy-related comorbidities such as cognitive deficits and mental health issues and integrating social support systems to improve the overall quality of life. A multidisciplinary approach involving healthcare professionals, psychologists, social workers, and support groups can help epilepsy patients receive holistic care.

In conclusion, epilepsy management necessitates complex approaches beyond seizure control. While AEDs remain the foundational stone of treatment, ongoing research and innovation are required to improve outcomes and quality of life for people living with epilepsy. Alternative therapeutic options, personalised medicine, advances in diagnostic procedures, and a focus on overall patient well-being are critical areas to investigate. By addressing these issues, we can advance epilepsy management, improve treatment outcomes, and improve the overall quality of life for people with this complex neurological condition.

8. Summary of the chapter

The chapter explored the potential therapeutic implications of oxytocin in epilepsy. Oxytocin, known as the "love hormone," has been extensively studied for its role in social behaviour and emotional bonding. Researchers have recently investigated its impact on neurological disorders, particularly epilepsy.

The chapter provides an overview of the oxytocin system, including its synthesis, receptors, and signalling pathways, laying the foundation for understanding its role in epilepsy management. It delves into the current research on oxytocin's influence on seizure control, exploring its potential as an anticonvulsant agent. Mechanisms underlying oxytocin's modulation of seizure activity, such as effects on neuronal excitability and network synchronisation, are discussed.

Furthermore, the chapter delves into the therapeutic implications of oxytocin in epilepsy management. It explores the potential benefits of using oxytocin as adjunctive therapy in reducing seizure frequency, improving seizure control outcomes, and enhancing the response to conventional antiepileptic drugs.

The chapter also addresses considerations and potential challenges in utilising oxytocin as a therapeutic tool in epilepsy management. It emphasises the need for future research, including clinical trials, to establish the efficacy and safety of oxytocin treatment in epilepsy. Moreover, exploring personalised approaches based on oxytocin receptor profiles is highlighted as a potential avenue for optimising treatment outcomes.

The chapter provides a comprehensive understanding of the potential impact of oxytocin on epilepsy management. Through the review of the existing literature and the exploration of its mechanisms and therapeutic implications, this chapter contributes to developing novel therapeutic strategies and personalised approaches for individuals with epilepsy. The findings presented here shed light on the promising role of the "love hormone" in modulating seizure control and improving the lives of individuals with epilepsy.

Conflict of interest

The authors declare no conflict of interest.

Author details

Lufuno Makhado* and Thendo Gertie Makhado
University of Venda, Thohoyandou, South Africa

*Address all correspondence to: lufuno.makhado@univen.ac.za

References

[1] Winter AL, Granqvist P. Where the Spirit meets the bone: Embodied Religiospiritual cognition from an attachment viewpoint. Religions. 2023;**14**(4):511

[2] Floricica CM, Luminita SM, Roxana NV. The importance of breastfeeding on the development of the mother-child relationship from an emotional point of view. Technium Social Sciences Journal. 2021;**23**:467

[3] Kovács B, Kéri S. Off-label intranasal oxytocin use in adults is associated with increased amygdala-cingulate resting-state connectivity. European Psychiatry. 2015;**30**(4):542-547

[4] Olff M. Bonding after trauma: On the role of social support and the oxytocin system in traumatic stress. European Journal of Psychotraumatology. 2012;**3**(1):18597

[5] Feldman R. Oxytocin and social affiliation in humans. Hormones and Behavior. 2012;**61**(3):380-391

[6] Zak PJ. The neuroscience of high-trust organizations. Consulting Psychology Journal: Practice and Research. 2018;**70**(1):45

[7] Yamasue H. Function and structure in social brain regions can link oxytocin-receptor genes with autistic social behavior. Brain Dev. 2013;**35**(2):111-118

[8] Kamp-Becker I, Poustka L, Bachmann C, Ehrlich S, Hoffmann F, Kanske P, et al. Study protocol of the ASD-net, the German research consortium for the study of autism Spectrum disorder across the lifespan: From a better etiological understanding, through valid diagnosis, to more effective health care. BMC Psychiatry. 2017;**17**(1):1-4

[9] Mayer AV, Wermter AK, Stroth S, Alter P, Haberhausen M, Stehr T, et al. Randomized clinical trial shows no substantial modulation of empathy-related neural activation by intranasal oxytocin in autism. Scientific Reports. 2021;**11**(1):15056

[10] Chen Z, Brodie MJ, Liew D, Kwan P. Treatment outcomes in patients with newly diagnosed epilepsy treated with established and new antiepileptic drugs: A 30-year longitudinal cohort study. JAMA Neurology. 2018;**75**(3):279-286

[11] Kwan P, Schachter SC, Brodie MJ. Drug-resistant epilepsy. The New England Journal of Medicine. 2011;**365**:919-926

[12] Meyer-Lindenberg A, Domes G, Kirsch P, Heinrichs M. Oxytocin and vasopressin in the human brain: Social neuropeptides for translational medicine. Nature Reviews Neuroscience. 2011;**12**(9):524-538

[13] Lin YT, Hsu KS. Oxytocin receptor signaling in the hippocampus: Role in regulating neuronal excitability, network oscillatory activity, synaptic plasticity and social memory. Progress in Neurobiology. 2018;**171**:1-4

[14] Lapp HE, Bartlett AA, Zup SL, Hunter RG, Moore CL. Early experience alters developmental trajectory of central oxytocin systems involved in hypothalamic-pituitary-adrenal axis regulation in long-Evans rats. Hormones and Behavior. 2020;**126**:104822

[15] Grebe NM, Gangestad SW. An evolutionary framework. In: The Oxford

Handbook of Evolutionary Psychology and Behavioral Endocrinology. 2019. p. 317

[16] Stolzenberg DS, Hernandez-D'Anna KL, Bosch OJ, Lonstein JS. Maternal behavior from a neuroendocrine perspective. In: Oxford Research Encyclopedia of Neuroscience. United Kingdom; 2019

[17] Smith JA, Eikenberry SA, Scott KA, Baumer-Harrison C, de Lartigue G, de Kloet AD, et al. Oxytocin and cardiometabolic interoception: Knowing oneself affects ingestive and social behaviors. Appetite. 2022;**175**:106054

[18] Ma Y, Shamay-Tsoory S, Han S, Zink CF. Oxytocin and social adaptation: Insights from neuroimaging studies of healthy and clinical populations. Trends in Cognitive Sciences. 2016;**20**(2):133-145

[19] Quattrocki E, Friston K. Autism, oxytocin and interoception. Neuroscience & Biobehavioral Reviews. 2014;**47**:410-430

[20] MacDonald K, MacDonald TM. The peptide that binds: A systematic review of oxytocin and its prosocial effects in humans. Harvard Review of Psychiatry. 2010;**18**(1):1-21

[21] Yamasue H, Yee JR, Hurlemann R, Rilling JK, Chen FS, Meyer-Lindenberg A, et al. Integrative approaches utilizing oxytocin to enhance prosocial behavior: From animal and human social behavior to autistic social dysfunction. Journal of Neuroscience. 2012;**32**(41):14109-1417a

[22] Striepens N, Kendrick KM, Maier W, Hurlemann R. Prosocial effects of oxytocin and clinical evidence for its therapeutic potential. Frontiers in Neuroendocrinology. 2011;**32**(4):426-450

[23] Campbell A. Oxytocin and human social behavior. Personality and Social Psychology Review. 2010;**14**(3):281-295

[24] De Dreu CK. Oxytocin modulates cooperation within and competition between groups: An integrative review and research agenda. Hormones and Behavior. 2012;**61**(3):419-428

[25] Olff M, Frijling JL, Kubzansky LD, Bradley B, Ellenbogen MA, Cardoso C, et al. The role of oxytocin in social bonding, stress regulation and mental health: An update on the moderating effects of context and interindividual differences. Psychoneuroendocrinology. 2013;**38**(9):1883-1894

[26] Davis MC, Green MF, Lee J, Horan WP, Senturk D, Clarke AD, et al. Oxytocin-augmented social cognitive skills training in schizophrenia. Neuropsychopharmacology. 2014;**39**(9):2070-2077

[27] Graustella AJ, MacLeod C. A critical review of the influence of oxytocin nasal spray on social cognition in humans: Evidence and future directions. Hormones and Behavior. 2012;**61**(3):410-418

[28] Ebert A, Brüne M. Oxytocin and social cognition. Behavioral Pharmacology of Neuropeptides: Oxytocin. 2018;**35**:375-388

[29] Ellenbogen MA. Oxytocin and facial emotion recognition. Behavioral Pharmacology of Neuropeptides: Oxytocin. 2018;**35**:349-374

[30] Bartz JA, Zaki J, Bolger N, Ochsner KN. Social effects of oxytocin in humans: Context and person matter. Trends in Cognitive Sciences. 2011;**15**(7):301-309

[31] Domes G, Heinrichs M, Gläscher J, Büchel C, Braus DF, Herpertz SC. Oxytocin attenuates amygdala responses to emotional faces regardless of valence. Biological Psychiatry. 2007;**62**(10):1187-1190

[32] Spilka MJ, Keller WR, Buchanan RW, Gold JM, Koenig JI, Strauss GP. Endogenous oxytocin levels are associated with facial emotion recognition accuracy but not gaze behavior in individuals with schizophrenia. Acta Psychiatrica Scandinavica. 2022;**145**(5):494-506

[33] Zhu S, Qing Y, Zhang Y, Zhang X, Ding F, Zhang R, et al. Transcutaneous auricular vagus nerve stimulation increases eye-gaze on salient facial features and oxytocin release. Psychophysiology. 2022;**59**(11):e14107

[34] Vehlen A, Kellner A, Normann C, Heinrichs M, Domes G. Reduced eye gaze during facial emotion recognition in chronic depression: Effects of intranasal oxytocin. Journal of Psychiatric Research. 2023;**159**:50-56

[35] Mesraoua B, Deleu D, Kullmann DM, Shetty AK, Boon P, Perucca E, et al. Novel therapies for epilepsy in the pipeline. Epilepsy & Behavior. 2019;**97**:282-290

[36] Kumar G. Evaluation and management of drug resistant epilepsy in children. Current Problems in Pediatric and Adolescent Health Care. 2021;**51**(7):101035

[37] Fatemi SH, Aldinger KA, Ashwood P, Bauman ML, Blaha CD, Blatt GJ, et al. Consensus paper: Pathological role of the cerebellum in autism. The Cerebellum. 2012;**11**:777-807

[38] Lochhead JJ, Thorne RG. Intranasal delivery of biologics to the central nervous system. Advanced Drug Delivery Reviews. 2012;**64**(7):614-628

[39] Janković SM, Đešević M. Advancements in neuroactive peptides in seizures. Expert Review of Neurotherapeutics. 2022;**22**(2):129-143

[40] Clynen E, Swijsen A, Raijmakers M, Hoogland G, Rigo JM. Neuropeptides as targets for the development of anticonvulsant drugs. Molecular Neurobiology. 2014;**50**:626-646

[41] Raso GM, Russo R, Calignano A, Meli R. Palmitoylethanolamide in CNS health and disease. Pharmacological Research. 2014;**86**:32-41

[42] Kaur S, Singh S, Arora A, Ram P, Kumar S, Kumar P, et al. Pharmacology of GABA and its receptors. Frontiers in Pharmacology of Neurotransmitters. 2020;**16**:241-292

[43] Aguilar-Raab C, Eckstein M, Geracitano S, Prevost M, Gold I, Heinrichs M, et al. Oxytocin modulates the cognitive appraisal of the own and others close intimate relationships. Frontiers in Neuroscience. 2019;**13**:714

[44] Eckstein M, Stößel G, Gerchen MF, Bilek E, Kirsch P, Ditzen B. Neural responses to instructed positive couple interaction: An fMRI study on compliment sharing. Social Cognitive and Affective Neuroscience. 2023;**18**(1):nsad005

[45] Xu L, Becker B, Kendrick KM. Oxytocin facilitates social learning by promoting conformity to trusted individuals. Frontiers in Neuroscience. 2019;**13**:56

[46] Schiller B, Brustkern J, Walker M, Hamm A, Heinrichs M. Oxytocin has sex-specific effects on trust and underlying neurophysiological processes. Psychoneuroendocrinology. 2023;**151**:106076

[47] Tarsha MS. Neurobiology of social capacities: The building

blocks of friendship. Peace Review. 2022;**34**(3):390-401

[48] Stevens F, Taber K. The neuroscience of empathy and compassion in pro-social behavior. Neuropsychologia. 2021;**159**:107925

[49] La Fratta I, Franceschelli S, Speranza L, Patruno A, Michetti C, D'Ercole P, et al. Salivary oxytocin, cognitive anxiety and self-confidence in pre-competition athletes. Scientific Reports. 2021;**11**(1):16877

[50] Elias LJ, Abdus-Saboor I. Bridging skin, brain, and behavior to understand pleasurable social touch. Current Opinion in Neurobiology. 2022;**73**:102527

[51] Grahn P, Ottosson J, Uvnäs-Moberg K. The oxytocinergic system as a mediator of anti-stress and instorative effects induced by nature: The calm and connection theory. Frontiers in Psychology. 2021;**12**:617814

[52] Sammut E, Sammut M, Chetcuti S, Calleja-Agius J. The role of oxytocin in maternal-fetal bonding and social interaction. International Journal of Prenatal and Life Sciences. 26 Aug 2023. DOI: 10.24946/IJPLS [Online]

[53] Nagasawa M, Kikusui T. Neuroendocrine mechanisms of social bonds and separation stress in rodents, dogs, and other species. In: Neuroscience of Social Stress. Cham: Springer International Publishing; 2021. pp. 3-22

[54] Patterson DK, Pollock D, Carter CS, Chambers JE. Treating opioid use disorder in peripartum mothers: A look at the psychodynamics, neurobiology, and potential role of oxytocin. Psychodynamic Psychiatry. 2021;**49**(1):48-72

[55] Uvnäs-Moberg K, Gross MM, Agius A, Downe S, Calleja-Agius J. Are there epigenetic oxytocin-mediated effects on the mother and infant during physiological childbirth? International Journal of Molecular Sciences. 2020;**21**(24):9503

[56] Yurtsal ZB, Uslu D. Sexual aspects of breastfeeding and lactation. In: Midwifery and Sexuality. Cham: Springer International Publishing; 2023. pp. 99-111

[57] Piccini O. The mother's body, the role of pleasure in the mother-infant relationship, and the traumatic risk. International Forum of Psychoanalysis. 2021;**30**(3):129-138

[58] Feldman R. What is resilience: An affiliative neuroscience approach. World Psychiatry. 2020;**19**(2):132-150

[59] Sharvin BL, Aburto MR, Cryan JF. Decoding the neurocircuitry of gut feelings: Region-specific microbiome-mediated brain alterations. Neurobiology of Disease. 2023;**179**:106033

[60] Gangopadhyay P, Chawla M, Dal Monte O, Chang SW. Prefrontal–amygdala circuits in social decision-making. Nature Neuroscience. 2021;**24**(1):5-18

[61] Inoue K, Ford CL, Horie K, Young LJ. Oxytocin receptors are widely distributed in the prairie vole (Microtus ochrogaster) brain: Relation to social behavior, genetic polymorphisms, and the dopamine system. Journal of Comparative Neurology. 2022;**530**(16):2881-2900

[62] Bhattarai JP, Etyemez S, Jaaro-Peled H, Janke E, Tolosa UD, Kamiya A, et al. Olfactory modulation of the medial prefrontal cortex circuitry: Implications for social cognition. Seminars in Cell & Developmental Biology. 2022;**129**:31-39. DOI: 10.1016/j.semcdb.2021.03.022

[63] Iovino M, Messana T, Tortora A, Giusti C, Lisco G, Giagulli VA, et al. Oxytocin signalling pathway: From cell biology to clinical implications. Endocrine, Metabolic & Immune Disorders-Drug Targets (Formerly Current Drug Targets-Immune, Endocrine & Metabolic Disorders). 2021;**21**(1):91-110

[64] Pierzynowska K, Gaffke L, Żabińska M, Cyske Z, Rintz E, Wiśniewska K, et al. Roles of the oxytocin receptor (OXTR) in human diseases. International Journal of Molecular Sciences. 2023;**24**(4):3887

[65] Talpo F, Kaur N, Biella G. Oxytocin in brain health and disease: How can it exert such pleiotropic neuromodulatory effects? Frontiers in Molecular Neuroscience. 2023;**16**:1215351

[66] Baudon A, Creusot EC, Althammer F, Schaaf CP, Charlet A. Emerging role of astrocytes in oxytocin-mediated control of neural circuits and brain functions. Progress in Neurobiology. 2022;**217**:102328

[67] Gordon I, Jack A, Pretzsch CM, Vander Wyk B, Leckman JF, Feldman R, et al. Intranasal oxytocin enhances connectivity in the neural circuitry supporting social motivation and social perception in children with autism. Scientific Reports. 2016;**6**(1):1-4

[68] Gordon I, Vander Wyk BC, Bennett RH, Cordeaux C, Lucas MV, Eilbott JA, et al. Oxytocin enhances brain function in children with autism. Proceedings of the National Academy of Sciences. 2013;**110**(52):20953-20958

[69] Love TM. Oxytocin, motivation and the role of dopamine. Pharmacology Biochemistry and Behavior. 2014;**119**:49-60

[70] Matthews GA, Tye KM. Neural mechanisms of social homeostasis. Annals of the New York Academy of Sciences. 2019;**1457**(1):5-25

[71] Prior NH, Bentz EJ, Ophir AG. Reciprocal processes of sensory perception and social bonding: An integrated social-sensory framework of social behavior. Genes, Brain and Behavior. 2022;**21**(3):e12781

[72] Kavaliers M, Ossenkopp KP, Tyson CD, Bishnoi IR, Choleris E. Social factors and the neurobiology of pathogen avoidance. Biology Letters. 2022;**18**(2):20210371

[73] Yamasue H, Kuwabara H, Kawakubo Y, Kasai K. Oxytocin, sexually dimorphic features of the social brain, and autism. Psychiatry and Clinical Neurosciences. 2009;**63**(2):129-140

[74] Yoon S, Kim YK. The Role of the Oxytocin System in Anxiety Disorders. Anxiety Disorders: Rethinking and Understanding Recent Discoveries; 2020. pp. 103-120

[75] Coccia G, La Greca F, Di Luca M, Scheggia D. Dissecting social decision-making: A spotlight on oxytocinergic transmission. Frontiers in Molecular Neuroscience. 2022;**15**:1061934

[76] Perez-Rodriguez MM, Mahon K, Russo M, Ungar AK, Burdick KE. Oxytocin and social cognition in affective and psychotic disorders. European Neuropsychopharmacology. 2015;**25**(2):265-282

[77] Jankowski KF, Takahashi H. Cognitive neuroscience of social emotions and implications for psychopathology: Examining embarrassment, guilt, envy, and schadenfreude. Psychiatry and Clinical Neurosciences. 2014;**68**(5):319-336

[78] World Health Organization. Epilepsy. 2019. Available from: https://www.who.int/news-room/fact-sheets/detail/epilepsy

[79] Fisher RS, Cross JH, French JA, Higurashi N, Hirsch E, Jansen FE, et al. Operational classification of seizure types by the international league against epilepsy: Position paper of the ILAE commission for classification and terminology. Epilepsia. 2017;**58**(4):522-530

[80] Solanki C, Williams J, Wu C, Andrews C, Fayed I. Insula in epilepsy-"untying the gordian knot": A systematic review. Seizure. 2023;**106**:148-161

[81] Bomela W, Wang S, Chou CA, Li JS. Real-time inference and detection of disruptive EEG networks for epileptic seizures. Scientific Reports. 2020;**10**(1):8653

[82] Aftab M, Shah SA, Aslam AR, Saadeh W, Altaf MA. Design of energy-efficient electrocorticography recording system for intractable epilepsy in implantable environments. In: 2020 IEEE International Symposium on Circuits and Systems (ISCAS), Seville, Spain. 2020. pp. 1-5. DOI: 10.1109/ISCAS45731.2020.9180498

[83] Ozsoy HZ. Anticonvulsant effects of carbonic anhydrase inhibitors: The enigmatic link between carbonic anhydrases and electrical activity of the brain. Neurochemical Research. 2021;**46**(11):2783-2799

[84] Milligan TA. Epilepsy: A clinical overview. The American Journal of Medicine. 2021;**134**(7):840-847

[85] Beniczky S, Tatum WO, Blumenfeld H, Stefan H, Mani J, Maillard L, et al. Seizure semiology: ILAE glossary of terms and their significance. Epileptic Disorders. 2022;**24**(3):447-495

[86] Surges R, Shmuely S, Dietze C, Ryvlin P, Thijs RD. Identifying patients with epilepsy at high risk of cardiac death: Signs, risk factors and initial management of high risk of cardiac death. Epileptic Disorders. 2021;**23**(1):17-39

[87] Dhondiyal SA, Dimri SC. Diagnosis, treatment and containment of epilepsy disease. In: 2022 Seventh International Conference on Parallel, Distributed and Grid Computing (PDGC), Solan, Himachal Pradesh, India. 2022. pp. 269-275. DOI: 10.1109/PDGC56933.2022.10053220

[88] Campbell TP, Kelly KM, Quezada TA. Handbook of Emergency Neurology. United Kingdom: Cambridge; 2023. p. 189

[89] Van Loo KM, Carvill GL, Becker AJ, Conboy K, Goldman AM, Kobow K, et al. Epigenetic genes and epilepsy—Emerging mechanisms and clinical applications. Nature Reviews Neurology. 2022;**18**(9):530-543

[90] Thakran S, Guin D, Singh P, Singh P, Kukal S, Rawat C, et al. Genetic landscape of common epilepsies: Advancing towards precision in treatment. International Journal of Molecular Sciences. 2020;**21**(20):7784

[91] Sun Y, Lu L, Li L, Wang J. An advance about the genetic causes of epilepsy. In: E3S Web of Conferences. Vol. 271. EDP Sciences; 2021. p. 03068

[92] Ramos-Riera KP, Pérez-Severiano F, López-Meraz ML. Oxidative stress: A common imbalance in diabetes and epilepsy. Metabolic Brain Disease. 2023;**38**(3):767-782

[93] Grove RA, Madhavan D, Boone CH, Braga CP, Papackova Z, Kyllo H, et al. Aberrant energy metabolism and redox balance in seizure onset zones of

epileptic patients. Journal of Proteomics. 2020;**223**:103812

[94] Bartolini E, Ferrari AR, Fiori S, Della VS. Glycaemic imbalances in seizures and epilepsy of Paediatric age: A literature review. Journal of Clinical Medicine. 2023;**12**(7):2580

[95] Pack AM. Epilepsy overview and revised classification of seizures and epilepsies. CONTINUUM: Lifelong learning. Neurology. 2019;**25**(2):306-321

[96] Egesa IJ, Newton CR, Kariuki SM. Evaluation of the international league against epilepsy 1981, 1989, and 2017 classifications of seizure semiology and etiology in a population-based cohort of children and adults with epilepsy. Epilepsia Open. 2022;**7**(1):98-109

[97] Thangarajan R, Bakthavatchalam P. Iron and epilepsy. In: Brain-Iron Cross Talk. Singapore: Springer Nature Singapore; 2022. pp. 189-209

[98] Swanson KI, Smith KA, Mirzadeh Z, Ponce FA. Epilepsy, functional neurosurgery, and pain. Operative Neurosurgery. 2019;**17**(Supplement_2):S209-S228

[99] Milton J, Jung P, editors. Epilepsy as a Dynamic Disease. Germany: Springer Science & Business Media; 2003

[100] Marsh L, Rao V. Psychiatric complications in patients with epilepsy: A review. Epilepsy Research. 2002;**49**(1):11-33

[101] Ahmed M, Nasir M, Yalew S, Getahun F, Getahun F. Assessment of treatment outcome and its associated factors among adult epileptic patients in public hospitals in the southern Ethiopia: A multi-center cross-sectional study. Ethiopian Journal of Health Sciences. 2023;**33**(2):327-336. DOI: 10.4314/ejhs.v33i2.18

[102] Kim H, Kim DW, Lee ST, Byun JI, Seo JG, No YJ, et al. Antiepileptic drug selection according to seizure type in adult patients with epilepsy. Journal of Clinical Neurology (Seoul, Korea). 2020;**16**(4):547

[103] Padda J, Khalid K, Syam M, Kakani V, Tonpouwo GK, Dhakal R, et al. Association of anemia with epilepsy and antiepileptic drugs. Cureus. 2021;**13**(11):e19334. DOI: 10.7759/cureus.19334

[104] Łukawski K, Czuczwar SJ. Emerging therapeutic targets for epilepsy: Preclinical insights. Expert Opinion on Therapeutic Targets. 2022;**26**(3):193-206

[105] Elger CE, Schmidt D. Modern management of epilepsy: A practical approach. Epilepsy & Behavior. 2008;**12**(4):501-539

[106] Kneen R, Appleton RE. Alternative approaches to conventional antiepileptic drugs in the management of paediatric epilepsy. Archives of Disease in Childhood. 2006;**91**(11):936-941

[107] Kobau R, DiIorio C. Epilepsy self-management: A comparison of self-efficacy and outcome expectancy for medication adherence and lifestyle behaviors among people living with epilepsy. Epilepsy & Behavior. 2003;**4**(3):217-225

[108] Xue T, Chen S, Bai Y, Han C, Yang A, Zhang J. Neuromodulation in drug-resistant epilepsy: A review of current knowledge. Acta Neurologica Scandinavica. 2022;**146**(6):786-797

[109] Toffa DH, Touma L, El Meskine T, Bouthillier A, Nguyen DK. Learnings from 30 years of reported efficacy and safety of vagus nerve stimulation (VNS) for epilepsy treatment: A critical review. Seizure. 2020;**83**:104-123

[110] Austelle CW, O'Leary GH, Thompson S, Gruber E, Kahn A, Manett AJ, et al. A comprehensive review of vagus nerve stimulation for depression. Neuromodulation: Technology at the neural. Interface. 2022;**25**(3):309-315

[111] Lin Y, Wang Y. Neurostimulation as a promising epilepsy therapy. Epilepsia Open. 2017;**2**(4):371-387

[112] Davis P, Gaitanis J. Neuromodulation for the treatment of epilepsy: A review of current approaches and future directions. Clinical Therapeutics. 2020;**42**(7):1140-1154

[113] Kossoff EH, Zupec-Kania BA, Auvin S, Ballaban-Gil KR, Christina Bergqvist AG, Blackford R, et al. Optimal clinical management of children receiving dietary therapies for epilepsy: Updated recommendations of the international ketogenic diet study group. Epilepsia Open. 2018;**3**(2):175-192

[114] Cervenka MC, Kossoff EH. Dietary treatment of intractable epilepsy. Continuum: Lifelong Learning in Neurology. 2013;**19**(3):756-766

[115] Kossoff EH, Wang HS. Dietary therapies for epilepsy. Biomed Journal. 2013;**36**(1):2-8

[116] Hartman AL, Vining EP. Clinical aspects of the ketogenic diet. Epilepsia. 2007;**48**(1):31-42

[117] Felton EA, Cervenka MC. Dietary therapy is the best option for refractory nonsurgical epilepsy. Epilepsia. 2015;**56**(9):1325-1329

[118] Simpson HD, Schulze-Bonhage A, Cascino GD, Fisher RS, Jobst BC, Sperling MR, et al. Practical considerations in epilepsy neurostimulation. Epilepsia. 2022;**63**(10):2445-2460

[119] Yan H, Ibrahim GM. Resective epilepsy surgery involving eloquent cortex in the age of responsive neurostimulation: A value-based decision-making framework. Epilepsy & Behavior. 2019;**99**:106479

[120] Khan M, Paktiawal J, Piper RJ, Chari A, Tisdall MM. Intracranial neuromodulation with deep brain stimulation and responsive neurostimulation in children with drug-resistant epilepsy: A systematic review. Journal of Neurosurgery: Pediatrics. 2021;**29**(2):208-217

[121] Jarosiewicz B, Morrell M. The RNS system: Brain-responsive neurostimulation for the treatment of epilepsy. Expert Review of Medical Devices. 2021;**18**(2):129-138

[122] Roa JA, Marcuse L, Fields M, La Vega-Talbott M, Yoo JY, Wolf SM, et al. Long-term outcomes after responsive neurostimulation for treatment of refractory epilepsy: A single-center experience of 100 cases. Journal of Neurosurgery. 2023;**1**(aop):1-8

[123] Jobst BC, Cascino GD. Resective epilepsy surgery for drug-resistant focal epilepsy: A review. Journal of the American Medical Association. 2015;**313**(3):285-293. DOI: 10.1001/jama.2014.17426

[124] Sone D. Neurobiological mechanisms of psychosis in epilepsy: Findings from neuroimaging studies. Frontiers in Psychiatry. 2022;**13**:1079295

[125] Bonakdarpour B, Takarabe C. Brain networks, clinical manifestations, and neuroimaging of cognitive disorders: The role of computed tomography (CT), magnetic resonance imaging (MRI), positron emission tomography (PET), and other advanced neuroimaging tests. Clinics in Geriatric Medicine. 2023;**39**(1):45-65

[126] Goodman AM, Szaflarski JP. Recent advances in neuroimaging of epilepsy. Neurotherapeutics. 2021;**18**(2):811-826

[127] Aceves-Serrano L, Neva JL, Doudet DJ. Insight into the effects of clinical repetitive transcranial magnetic stimulation on the brain from positron emission tomography and magnetic resonance imaging studies: A narrative review. Frontiers in Neuroscience. 2022;**16**:787403

[128] Knowles JK, Helbig I, Metcalf CS, Lubbers LS, Isom LL, Demarest S, et al. Precision medicine for genetic epilepsy on the horizon: Recent advances, present challenges, and suggestions for continued progress. Epilepsia. 2022;**63**(10):2461-2475

[129] Striano P, Minassian BA. From genetic testing to precision medicine in epilepsy. Neurotherapeutics. 2020;**17**(2):609-615

[130] Scala M, Bianchi A, Bisulli F, Coppola A, Elia M, Trivisano M, et al. Advances in genetic testing and optimization of clinical management in children and adults with epilepsy. Expert Review of Neurotherapeutics. 2020;**20**(3):251-269

[131] Hopkins CE, Brock T, Caulfield TR, Bainbridge M. Phenotypic screening models for rapid diagnosis of genetic variants and discovery of personalized therapeutics. Molecular Aspects of Medicine. 2022;**91**:101153

[132] Møller RS, Larsen LH, Johannesen KM, Talvik I, Talvik T, Vaher U, et al. Gene panel testing in epileptic encephalopathies and familial epilepsies. Molecular Syndromology. 2016;**7**(4):210-219

[133] Mizuno S, Koneval Z, Zierath DK, Knox KM, White HS, Barker-Haliski M. Diurnal burden of spontaneous seizures in early epileptogenesis in the post-kainic acid rat model of epilepsy. Epilepsia Open. 2021;**6**(2):431-436

[134] Beghi E. The epidemiology of epilepsy. Neuroepidemiology. 2020;**54**(2):185-191

[135] Omidvar M, Zahedi A, Bakhshi H. EEG signal processing for epilepsy seizure detection using 5-level Db4 discrete wavelet transform, GA-based feature selection and ANN/SVM classifiers. Journal of Ambient Intelligence and Humanized Computing. 2021;**12**:10395-10403

[136] Kumar TR, Geetha K, Remmiya Devi G, Barkath NS. Process efficient artificial neural network-based approach for channel selection and classification of seizures. In: Computational Network Application Tools for Performance Management. 2020. pp. 119-125

[137] Bartolomei F, Blumenfeld H. Epilepsy and consciousness. In: Neurobiology of the Epilepsies: From Epilepsy: A Comprehensive Textbook. 2022

[138] Leibetseder A, Eisermann M, LaFrance WC Jr, Nobili L, von Oertzen TJ. How to distinguish seizures from non-epileptic manifestations. Epileptic Disorders. 2020;**22**(6):716-738

[139] Balestrini S, Arzimanoglou A, Blümcke I, Scheffer IE, Wiebe S, Zelano J, et al. The aetiologies of epilepsy. Epileptic Disorders. 2021;**23**(1):1-6

[140] Bernardo P, Cinalli G, Santoro C. Epilepsy in NF1: A systematic review of the literature. Child's Nervous System. 2020;**36**:2333-2350

[141] Ramos AP, Burneo JG. Seizures and epilepsy associated with central nervous system tuberculosis. Seizure. 2023;**107**:60-66

[142] Mauritz M, Hirsch LJ, Camfield P, Chin R, Nardone R, Lattanzi S, et al. Acute symptomatic

seizures: An educational, evidence-based review. Epileptic Disorders. 2022;**24**(1):26-49

[143] Sandoval Karamian AG, Wusthoff CJ, Boothroyd D, Yeom KW, Knowles JK. Neonatal genetic epilepsies display convergent white matter microstructural abnormalities. Epilepsia. 2020;**61**(12):e192-e197

[144] von Stülpnagel C, Van Baalen A, Borggraefe I, Eschermann K, Hartlieb T, Kiwull L, et al. Network for therapy in rare epilepsies (NETRE): Lessons from the past 15 years. Frontiers in Neurology. 2021;**11**:622510

[145] Milh M, Riccardi F, Denis J. Genetics of neonatal onset epilepsies: An overview. Revue Neurologique. 2020;**176**(1-2):2-9

[146] Sawires R, Buttery J, Fahey M. A review of febrile seizures: Recent advances in understanding of febrile seizure pathophysiology and commonly implicated viral triggers. Frontiers in Pediatrics. 2022;**9**:1596

[147] Perucca P, Bahlo M, Berkovic SF. The genetics of epilepsy. Annual Review of Genomics and Human Genetics. 2020;**21**:205-230

[148] Lu L, Xiong W, Liu D, Liu J, Yang D, Li N, et al. New onset acute symptomatic seizure and risk factors in coronavirus disease 2019: A retrospective multicenter study. Epilepsia. 2020;**61**(6):e49-e53

[149] Cheng JY, French JA. Intelligent use of antiepileptic drugs is beneficial to patients. Current Opinion in Neurology. 2018;**31**(2):169-175

[150] Kanner AM, Ashman E, Gloss D, Harden C, Bourgeois B, Bautista JF, et al. Practice guideline update summary: Efficacy and tolerability of the new antiepileptic drugs I: Treatment of new-onset epilepsy: Report of the guideline development, dissemination, and implementation Subcommittee of the American Academy of neurology and the American Epilepsy Society. Neurology. 2018;**91**(2):74-81

[151] Perucca P, Scheffer IE, Kiley M. The management of epilepsy in children and adults. Medical Journal of Australia. 2018;**208**(5):226-233

[152] Löscher W, Klein P. The pharmacology and clinical efficacy of antiseizure medications: From bromide salts to cenobamate and beyond. CNS Drugs. 2021;**35**(9):935-963

[153] Brückner KE. Cognitive and psychological side effects of antiepileptic drugs [Internet]. In: Epilepsy - Update on Classification, Etiologies, Instrumental Diagnosis and Treatment. London, UK: IntechOpen; 2021. DOI: 10.5772/intechopen.94308

[154] Engel J Jr. Evolution of concepts in epilepsy surgery. Epileptic Disorders. 2019;**21**(5):391-409

[155] Bäuerle P, Schneider U, Holtkamp M, Gloveli T, Dugladze T. Outlines to initiate epilepsy surgery in low-and middle-income countries. Journal of Integrative Neuroscience. 2022;**21**(5):134

[156] Sheikh S, Thompson N, Bingaman W, Gonzalez-Martinez J, Najm I, Jehi L. (Re) defining success in epilepsy surgery: The importance of relative seizure reduction in patient-reported quality of life. Epilepsia. 2019;**60**(10):2078-2085

[157] Tsai JD, Fan PC, Lee WT, Hung PL, Hung KL, Wang HS, et al. Vagus nerve stimulation in pediatric patients with failed epilepsy surgery. Acta Neurologica Belgica. 2021;**121**:1305-1309

[158] Sheikh SR, Nair D, Gross RE, Gonzalez-Martinez J. Tracking a changing paradigm and the modern face of epilepsy surgery: A comprehensive and critical review on the hunt for the optimal extent of resection in mesial temporal lobe epilepsy. Epilepsia. 2019;**60**(9):1768-1793

[159] Zou W, Guo Z, Suo L, Zhu J, He H, Li X, et al. Nucleus accumbens shell modulates seizure propagation in a mouse temporal lobe epilepsy model. Frontiers in Cell and Developmental Biology. 2022;**10**:1031872

[160] Gawel K, Langlois M, Martins T, van der Ent W, Tiraboschi E, Jacmin M, et al. Seizing the moment: Zebrafish epilepsy models. Neuroscience & Biobehavioral Reviews. 2020;**116**:1-20

[161] Javaid S, Alqahtani F, Ashraf W, Anjum SM, Rasool MF, Ahmad T, et al. Tiagabine suppresses pentylenetetrazole-induced seizures in mice and improves behavioral and cognitive parameters by modulating BDNF/TrkB expression and neuroinflammatory markers. Biomedicine & Pharmacotherapy. 2023;**160**:114406

[162] Li C, Kash TL. κ-Opioid receptor modulation of GABAergic inputs onto ventrolateral periaqueductal gray dopamine neurons. Complex Psychiatry. 2019;**5**(4):190-199

[163] Amrutkar DV, Dyhring T, Jacobsen TA, Larsen JS, Sandager-Nielsen K. Anti-tremor action of subtype selective positive allosteric modulators of GABA a receptors in a rat model of essential tremors. The Cerebellum. 2020;**19**:265-274

[164] Patel RR, Wolfe SA, Bajo M, Abeynaike S, Pahng A, Borgonetti V, et al. IL-10 normalizes aberrant amygdala GABA transmission and reverses anxiety-like behavior and dependence-induced escalation of alcohol intake. Progress in Neurobiology. 2021;**199**:101952

[165] Robinson SL, Perez-Heydrich CA, Thiele TE. Corticotropin releasing factor type 1 and 2 receptor signaling in the medial prefrontal cortex modulates binge-like ethanol consumption in C57BL/6J mice. Brain Sciences. 2019;**9**(7):171

[166] Riazi K, Galic MA, Kentner AC, Reid AY, Sharkey KA, Pittman QJ. Microglia-dependent alteration of glutamatergic synaptic transmission and plasticity in the hippocampus during peripheral inflammation. Journal of Neuroscience. 2015;**35**(12):4942-4952

[167] Chen Y, Xu J, Chen Y. Regulation of neurotransmitters by the gut microbiota and effects on cognition in neurological disorders. Nutrients. 2021;**13**(6):2099

[168] McCormack PL. Tranexamic acid: A review of its use in the treatment of hyperfibrinolysis. Drugs. 2012;**72**(5):585-617

[169] Wang Y, Zhan G, Cai Z, Jiao B, Zhao Y, Li S, et al. Vagus nerve stimulation in brain diseases: Therapeutic applications and biological mechanisms. Neuroscience & Biobehavioral Reviews. 2021;**127**:37-53

[170] Aruťuňanová P. Oxytocin and Social Behavior in Laboratory Rodents. Prague: Charles University; 2022

[171] Anagnostou E, Soorya L, Chaplin W, Bartz J, Halpern D, Wasserman S, et al. Intranasal oxytocin versus placebo in the treatment of adults with autism spectrum disorders: A randomized controlled trial. Molecular Autism. 2012;**3**(1):1-9

[172] Chen W, Man X, Zhang Y, Yao G, Chen J. Medial prefrontal cortex oxytocin mitigates epilepsy and

cognitive impairments induced by traumatic brain injury through reducing neuroinflammation in mice. Scientific Reports. 2023;**13**(1):5214

[173] Lu Y, Zhong R, Li M, Zhao Q, Zhang X, Hu B, et al. Social anxiety is associated with poor quality of life in adults with epilepsy in Northeast China: A cross-sectional study. Epilepsy & Behavior. 2021;**117**:107866

[174] Yao S, Chen Y, Zhuang Q, Zhang Y, Lan C, Zhu S, et al. Sniffing oxytocin: Nose to brain or nose to blood? Molecular Psychiatry. 2023:1-9. DOI: 10.1038/s41380-023-02075-2

Chapter 3

Social Cognitive Impairments as a Target of Non-Invasive Brain Stimulation for Functional Outcomes in Schizophrenia

Yuji Yamada and Tomiki Sumiyoshi

Abstract

Patients with schizophrenia suffer from impairments of social cognition that represent mental operations underlying real-world functioning. Pharmacological approaches have been attempted to overcome social cognitive disturbances of schizophrenia, but only yielded insufficient effects. As an alternative approach, some types of neuromodulations, particularly non-invasive brain stimulation, e.g., transcranial direct current stimulation (tDCS), have been drawing attention. While previous studies have performed anodal tDCS at the frontal brain regions, we hypothesized anodal stimulation at the temporal region would improve social cognitive function on the basis of the neural circuit governing it. Thus, our data indicate multisession tDCS delivered to the left superior temporal sulcus improves social cognition in patients with schizophrenia. In the present chapter, we overview studies of tDCS on social cognition and discuss optimal brain regions to be targeted for ameliorating symptoms and cognitive disturbances of schizophrenia.

Keywords: neuromodulation, transcranial direct current stimulation (tDCS), schizophrenia, social cognition, social function

1. Introduction

Schizophrenia is one of the most severe psychiatric diseases that affects approximately 0.75% of the world's population [1]. It is characterized by varying degrees of positive (e.g., auditory and/or visual hallucinations, delusions, psychomotor agitation, and paranoia), negative (e.g., apathy, deficits in motivation, social withdrawal, and reward-related functions), and cognitive symptoms (e.g., neurocognition and social cognition). Typically, the onset of schizophrenia is between the end of adolescence and the beginning of early adulthood. The disease has a chronic course with a myriad of psychotic episodes that generally lead to deterioration in real-world functional outcomes [2].

Cognitive impairment is a fundamental symptom of schizophrenia, with mild deficits appearing before the onset of psychosis, followed by an acute decline around

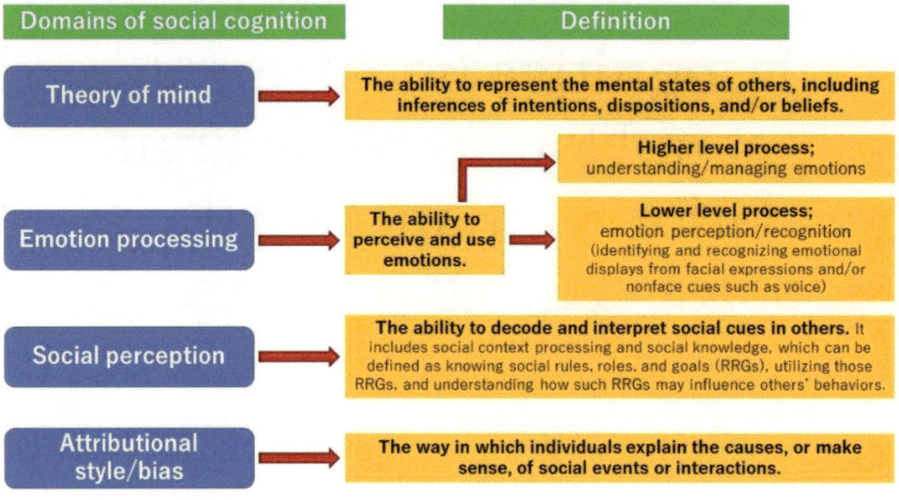

Figure 1.
Key domains of social cognition (adapted from Pinkham et al. [3]).

the first episode psychosis (FEP), and maintains to the chronic stages [2]. Specifically, neurocognition, e.g., working memory, verbal/visual learning and memory, attention, speed of processing, reasoning, and problem-solving, represents cognitive domains that are impaired in most patients. Similarly, impairments of social cognition have been considered as a target for the improvement of functional outcomes [2]. Social cognition is defined as mental operations underlying social behavior and includes domains of the theory of mind (ToM), emotion processing, social perception, and attributional bias/style (see **Figure 1**) [4]. Moreover, social cognition, especially ToM, is associated with social functioning [5], similar to the case for neurocognition [6]. Therefore, these areas of cognitive function have been considered a key target for interventions to improve the functionality of people with psychotic conditions.

Several pharmacological approaches have been tried to enhance social cognition in patients with schizophrenia [7]. In particular, intranasal oxytocin [8], psychostimulants (e.g., modafinil) [2], antipsychotic drugs (e.g., risperidone, olanzapine) [9], and acetylcholinesterase inhibitors [10] have been tested as candidate compounds. Of these, intranasal oxytocin and some atypical antipsychotic drugs may improve ToM [8] and emotion processing [9], respectively, while anti-dementia drugs (e.g., donepezil, galantamine, rivastigmine, and memantine) are less effective. Therefore, further research, including new therapeutic intervention techniques, is desperately needed to overcome social cognitive dysfunction.

2. Non-invasive brain stimulation for social cognitive impairments

As alternative approaches to the alleviation of social cognitive impairments, some types of neuromodulations, particularly non-invasive brain stimulation (NIBS), e.g., transcranial magnetic stimulation (TMS) or transcranial electrical stimulation (tES), have been attempted [11]. NIBS has the potential to improve functional outcomes by directly stimulating social brain areas, which is likely to facilitate or modulate neurotransmissions in the central nervous system [12, 13]. In particular, TMS has the advantage of accurately stimulating deeper cortical structures that are difficult to reach with tES. On the other hand, tES techniques are advantageous in terms of cost, portability, and safety [11]. Moreover, multi-session tES is thought to promote neuroplastic changes in cortical circuits by inducing long-term potentiation (LTP), which enhances the efficiency of information processing [12, 13]. Therefore, tES may improve functional outcomes by ameliorating social cognitive impairments in patients with schizophrenia.

Previous studies with TMS or tES, with the dorsolateral prefrontal cortex (DLPFC), as the stimulation site, have shown limited effects on social cognition, especially ToM [11]. For example, a single-session continuous theta burst stimulation, a type of TMS, over the right inferior parietal lobe (IPL) enhanced social perception [14], while multi-session high-frequency repetitive TMS over the left DLPFC improved emotional recognition (**Table 1**) [15]. On the other hand, single-session transcranial direct current stimulation (tDCS) delivered to the left frontal pole improved emotion recognition [16], whereas stimulation of the left DLPFC did not show such effects (**Table 2**) [17]. Overall, stimulations of frontal brain regions are advantageous to improve emotion recognition and social perception, but not ToM. To enhance the latter domain of social cognition, other stimulation sites need to be investigated.

The neural substrates governing social cognition include the superior temporal sulcus (STS), amygdala, medial prefrontal cortex, and orbitofrontal cortex. Specifically, functional connectivity among these brain regions is weakened in patients with schizophrenia [18–20]. Among them, the amygdala is involved in emotion recognition, while the prefrontal cortex is responsible for generating ToM. On the other hand, the STS is considered to play a role in both domains of social cognition (**Table 3**) [18, 19]. Therefore, we hypothesized that anodal stimulation over the left STS would be advantageous for treating social cognition disturbances [18, 19].

Study	**Walther et al. [14]**		**Wölwer et al. [15]**
Diagnosis	Schizophrenia		Schizophrenia
Sample size (active/ sham)	20/20		18/14
Location (stimulation)	Left IFG (iTBS)	Right IPL (cTBS)	Left DLPFC
Duration (days)	1		10
Evaluation	online		within 12 hours after stimulation
Outcomes	TULIA		Pictures of Facial Affect
Results	No significant effect	Significant effects on social perception	Significant effects in emotion recognition

IFG: Inferior Frontal Gyrus, IPL: Inferior Parietal Lobe, DLPFC: Dorsolateral Prefrontal Cortex, iTBS: intermittent Theta Burst Stimulation, cTBS: continuous Theta Burst Stimulation, TULIA: Test of Upper Limb Apraxia. In both studies [14, 15], sham stimulation was performed with a sham coil system without a magnetic field.

Table 1.
Characteristics of TMS studies (adapted from Yamada et al. [11]).

Study	Rassovsky et al., 2018 [17]	Rassovsky et al., 2015 [16]
Diagnosis	Schizophrenia	Schizophrenia
Sample size (active/sham)	37/37	12/12
Montage (Anode/cathode)	F3/Fp2	Fp1/Fp2
No. of sessions	2	1
Evaluation	online	online
Outcomes	TASIT, EAT, EIT, MSCEIT	TASIT, PONS, FEIT, MSCEIT
Results	No significant effect	Significant effects in emotion recognition

Montage is described by the International 10–20 electroencephalography system. TASIT: The Awareness of Social Inference Test, PONS: Profile of Nonverbal Sensitivity, EAT: Empathic Accuracy Task, EIT: Emotion Identification Test, FEIT: Facial Emotion Identification Test, MSCEIT: Mayer-Salovey-Caruso Emotional Intelligence Test.

Table 2.
Characteristics of tES studies (adapted from Yamada et al. [11]).

Domains of social cognition	Neural basis
Theory of mind (ToM)	Superior temporal sulcus (STS), medial prefrontal cortex (mPFC), middle temporal gyrus, etc.
Emotion recognition/processing	Amygdala, superior temporal sulcus (STS), medial prefrontal cortex (mPFC), inferior occipital gyrus, etc.
Attributional bias/style	Orbitofrontal cortex, insular cortex, striatum, amygdala, superior temporal sulcus (STS), etc.

Table 3.
Neural basis of social cognition (adapted from Yamada et al. [18]).

3. tDCS over the left temporal brain region improves social cognition

We conducted an open-label, single-arm trial designed to evaluate the effects and safety of multi-session tDCS on the left STS [18, 21]. The intervention was performed by a 1 × 1 transcranial direct current low-intensity stimulator (Model 1300 A; Soterix Medical Inc., New York, NY, USA). For the tDCS montage, the anode electrode was placed for the left STS and the cathode for the contralateral supraorbital region, which corresponded to the T3 and FP2 regions (**Figure 2**) in the International 10–20 electroencephalography system. We applied 10 sessions of direct current of 2 mA for 20 min on 5 consecutive days (twice per day, with an interval of 30 min). Clinical data were collected at baseline and 1 month after the final stimulus (**Table 4**). As a result, significant improvements were found on the tests of ToM, i.e., Social Cognition Screening Questionnaire (SCSQ) ($d = 0.53$) and Hinting Task scores ($d = 0.49$) (**Figure 3**) [21]. Moreover, serious adverse events were absent. To our knowledge, this study is the first to suggest the ability of multi-session tDCS delivered to the left STS to improve social cognition, especially ToM, in patients with schizophrenia.

Contrary to the results of our observations, previous studies report the absence of sufficient improvement of social cognitive functioning, including ToM, during or immediately after one or two sessions of tDCS in patients with schizophrenia (**Table 2**). By contrast, the effects in our study became evident 1 month after the last tDCS. The specific effects of multi-session tDCS that are associated with the

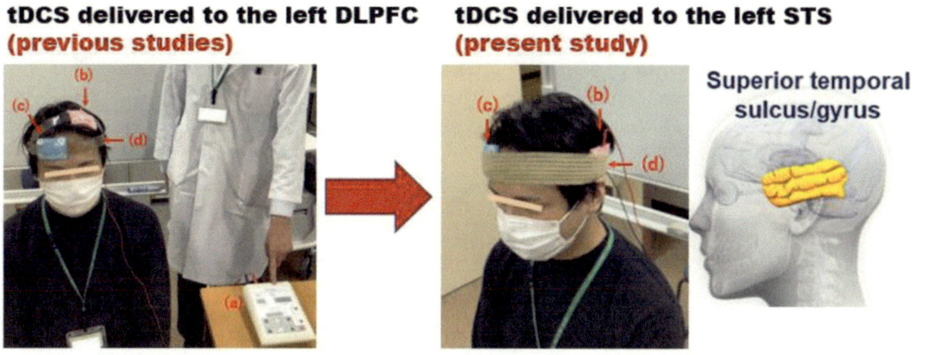

Figure 2.
Experimental setup for tDCS on the left DLPFC or STS (Yamada et al. [19]). tDCS: transcranial direct current stimulation, DLPFC: Dorsolateral Prefrontal Cortex, STS: Superior Temporal Sulcus. An administrator controls the stimulator (a). Anodal (b) and cathodal (c) electrodes of 35-cm² in size are placed on F3/T3 and the right supraorbital region, respectively. A head strap (d) is used as needed to increase reproducibility.

	Study period				
	Baseline	**Intervention**			**Follow-up**
Time point	Within 2 weeks before the start of intervention	Day 1	Days 2–4	Day 5	1 month after the end of the last stimulation
Enrollment					
Eligibility screen	X				
Informed consent	X				
Sociodemographic characteristics	X				
Intervention					
tDCS (twice/day)		X	X	X	
Assessments					
SCSQ	X				X
Hinting Task	X				X
FEST	X				X
Adverse events	X	X	X	X	X
Prescribed drugs	X	X	X	X	X

tDCS: transcranial direct current stimulation, SCSQ: Social Cognition Screening Questionnaire, FEST: Facial Emotion Selection Test. The timepoint of follow-up evaluation was allowed to be up to 7 days off.

Table 4.
Study schedule (adapted from Yamada et al. [18]).

prolonged (i.e., 1-month) change of social cognition may be mediated through several mechanisms, including the LTP, a continuous enhancement of signal transduction between neurons, and related neural events [12, 13]. As social cognition is considered to be linked to functional outcomes [5], multi-session, rather than single-session tDCS may provide benefits for functional recovery in patients with schizophrenia.

The current results, with tDCS as an interventional tool, may facilitate an understanding of the neural basis for social cognitive dysfunction. The neural circuity

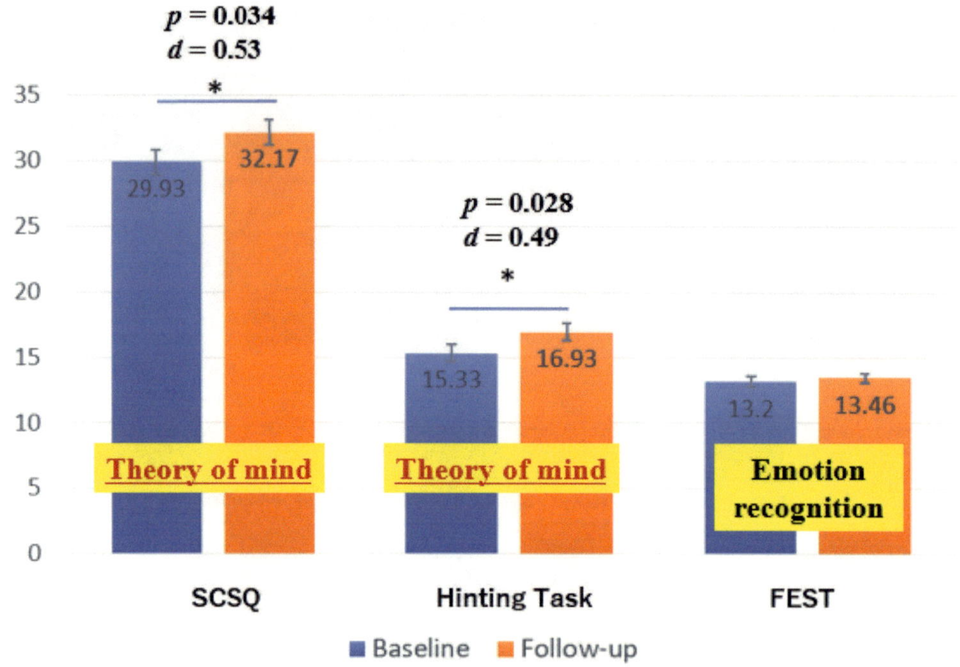

Figure 3.
Outcome measures at baseline and follow-up (1 month after the tDCS). tDCS: transcranial direct current stimulation, SCSQ: Social Cognition Screening Questionnaire, FEST: Facial Emotion Selection Test.

governing social cognition consists of the orbitofrontal cortex, medial prefrontal cortex, superior temporal sulcus/gyrus, and amygdala, in which the amygdala is considered as a hub [18–20]. Specifically, the amygdala interconnects with both temporal lobes, e.g., STS, and frontal lobes, e.g., orbitofrontal and medial prefrontal cortex (**Figure 4**) [20]. Also, the amygdala receives various sensory stimuli, and

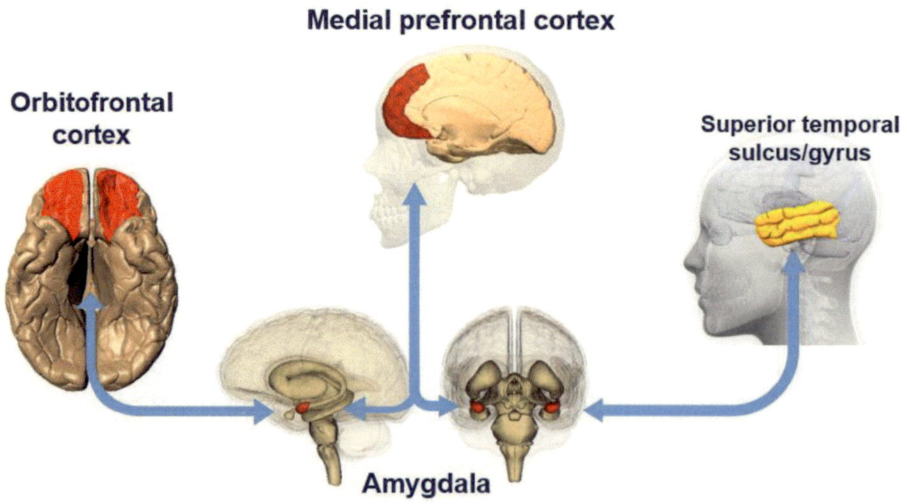

Figure 4.
Neural bases associated with social cognitive functions.

projects them to a wide range of cortical areas, most notably the STS and orbitofrontal cortex [20]. In patients with schizophrenia, functional connectivity among these brain regions is weakened [18–20]. Thus, pro-cognitive effects of tDCS may be associated with modification of functional connectivity, possibly through modulation of neurotransmissions [12, 13].

A strength of tDCS is its simplicity, which allows implementation in clinical settings without requiring expensive equipment, e.g., magnetic resonance imaging (MRI) machines. Such feasibility of tDCS is advantageous in, e.g., being built-in the telemedicine system.

4. Limitations

First, the small sample size of the present study may raise caution in assuming that the results can be generalized to the population. Second, as the study design is not randomized or blinded, the possibility of a placebo effect may not be completely ruled out. Third, although tDCS allows the selection of stimulation sites according to the targeted symptoms, it is difficult to fully ascertain if the electrical stimulation effectively modulates the targeted functional connectivity.

5. Conclusions

tDCS delivered to the left STS may produce benefits for some domains of social cognition, especially ToM, in patients with schizophrenia. Further investigations are warranted to determine if the type of functional outcome (e.g., daily-living activities, social function) would depend on the region of the brain stimulated by tDCS. These efforts may provide an effective therapeutic strategy to facilitate personal recovery for these patients.

Acknowledgements

We would like to thank Dr. Zui Narita and Mr. Kazuki Sueyoshi at the National Center of Neurology and Psychiatry for their valuable help.

Funding

The studies discussed in this chapter were partly supported by the Japan Society for the Promotion of Science KAKENHI Grant No.20 K16635 to YY, as well as JSPS KAKENHI Grant No. 20H03610, the Japan Health Research Promotion Bureau Grants (2020-B-08, 2021-B-01), Intramural Research Grant (2-3, 3-1) for Neurological and Psychiatric Disorders of the National Center of Neurology and Psychiatry to TS.

Conflict of interest

The authors declare no conflict of interest.

Ethics statement

The study was performed according to the Declaration of Helsinki and followed the Clinical Trials Act in Japan. The protocol was presented for approval by the National Center of Neurology and Psychiatry Clinical Research Review Board (CRB3180006).

Author details

Yuji Yamada[1] and Tomiki Sumiyoshi[2*]

1 Department of Psychiatry, National Center Hospital, National Center of Neurology and Psychiatry, Tokyo, Japan

2 Department of Preventive Intervention for Psychiatric Disorders, National Institute of Mental Health, National Center of Neurology and Psychiatry, Tokyo, Japan

*Address all correspondence to: sumiyot@ncnp.go.jp

References

[1] Moreno-Küstner B, Martín C, Pastor L. Prevalence of psychotic disorders and its association with methodological issues. A systematic review and meta-analyses. PLoS One. 2018;**13**(4):e0195687

[2] Yamada Y, Inagawa T, Sueyoshi K, Sugawara N, Ueda N, Omachi Y, et al. Social cognition deficits as a target of early intervention for psychoses: A systematic review. Frontiers in Psychiatry. 2019;**10**:333

[3] Pinkham AE, Penn DL, Green MF, Buck B, Healey K, Harvey PD. The social cognition psychometric evaluation study: Results of the expert survey and RAND panel. Schizophrenia Bulletin. 2014;**40**(4):813-823

[4] Couture SM, Penn DL, Roberts DL. The functional significance of social cognition in schizophrenia: A review. Schizophrenia Bulletin. 2006;**32**(Suppl 1): S44-S63

[5] Fett AK, Viechtbauer W, Dominguez MD, Penn DL, van Os J, Krabbendam L. The relationship between neurocognition and social cognition with functional outcomes in schizophrenia: A meta-analysis. Neuroscience and Biobehavioral Reviews. 2011;**35**(3):573-588

[6] Sumiyoshi T, Nishida K, Niimura H, Toyomaki A, Morimoto T, Tani M, et al. Cognitive insight and functional outcome in schizophrenia; a multi-center collaborative study with the specific level of functioning scale-Japanese version. Schizophraphy Research Cognition. 2016;**6**:9-14

[7] Yamada Y, Okubo R, Tachimori H, Uchino T, Kubota R, Okano H, et al. Pharmacological interventions for social cognitive impairments in schizophrenia: A protocol for a systematic review and network meta-analysis. Frontiers in Psychology. 2022;**13**:878829

[8] Bürkner PC, Williams DR, Simmons TC, Woolley JD. Intranasal oxytocin may improve high-level social cognition in schizophrenia, but not social cognition or Neurocognition in general: A multilevel Bayesian Meta-analysis. Schizophrenia Bulletin. 2017;**43**(6):1291-1303

[9] Gabay AS, Kempton MJ, Mehta MA. Facial affect processing deficits in schizophrenia: A meta-analysis of antipsychotic treatment effects. Journal of Psychopharmacology. 2015;**29**(2):224-229

[10] Kishi T, Ikuta T, Oya K, Matsunaga S, Matsuda Y, Iwata N. Anti-dementia drugs for psychopathology and cognitive impairment in schizophrenia: A systematic review and Meta-analysis. The International Journal of Neuropsychopharmacology. 2018;**21**(8):748-757

[11] Yamada Y, Inagawa T, Hirabayashi N, Sumiyoshi T. Emotion recognition deficits in psychiatric disorders as a target of non-invasive Neuromodulation: A systematic review. Clinical EEG and Neuroscience. 2022;**53**(6):506-512

[12] Yamada Y, Sumiyoshi T. Neurobiological mechanisms of transcranial direct current stimulation for psychiatric disorders; neurophysiological, chemical, and anatomical considerations. Frontiers in Human Neuroscience. 2021;**15**:631838

[13] Yamada Y, Sumiyoshi T. Preclinical evidence for the mechanisms

of transcranial direct current stimulation in the treatment of psychiatric disorders: A systematic review. Clinical EEG Neuroscience. 2021;**13**:15500594211066151

[14] Walther S, Kunz M, Müller M, Zürcher C, Vladimirova I, Bachofner H, et al. Single session transcranial magnetic stimulation ameliorates hand gesture deficits in schizophrenia. Schizophrenia Bulletin. 2020;**46**(2):286-293

[15] Wölwer W, Lowe A, Brinkmeyer J, Streit M, Habakuck M, Agelink MW, et al. Repetitive transcranial magnetic stimulation (rTMS) improves facial affect recognition in schizophrenia. Brain Stimulation. 2014;**7**(4):559-563

[16] Rassovsky Y, Dunn W, Wynn J, Wu AD, Iacoboni M, Hellemann G, et al. The effect of transcranial direct current stimulation on social cognition in schizophrenia: A preliminary study. Schizophrenia Research. 2015;**165**(2-3):171-174

[17] Rassovsky Y, Dunn W, Wynn JK, Wu AD, Iacoboni M, Hellemann G, et al. Single transcranial direct current stimulation in schizophrenia: Randomized, cross-over study of neurocognition, social cognition, ERPs, and side effects. PLoS One. 2018;**13**(5):e0197023

[18] Yamada Y, Inagawa T, Yokoi Y, Shirama A, Sueyoshi K, Wada A, et al. Efficacy and safety of multi-session transcranial direct current stimulation on social cognition in schizophrenia: A study protocol for an open-label, single-arm trial. Journal of Persian Medicine. 2021;**11**(4):317

[19] Yamada Y, Sumiyoshi T. Transcranial direct current stimulation and social cognition impairments of Schizophrenia; current knowledge and future Perspectives. Horizons in Neuroscience Research. 2022;**46**. New York: Nova Science Publishers:143-170

[20] Yamada Y, Matsumoto M, Iijima K, Sumiyoshi T. Specificity and continuity of schizophrenia and bipolar disorder: Relation to biomarkers. Current Pharmaceutical Design. 2020;**26**(2):191-200

[21] Yamada Y, Sueyoshi K, Yokoi Y, Inagawa T, Hirabayashi N, Oi H, et al. Transcranial direct current stimulation on the left superior temporal sulcus improves social cognition in schizophrenia: An open-label study. Frontiers in Psychiatry. 2022;**13**:862814

Chapter 4

Regulation of Oxytocin on Empathy and Its Neural Mechanism

Guangxin Yue

Abstract

Empathy is a multidimensional concept, including emotion and cognition. It plays a vital role in social communication, and it is very important for establishing harmonious relationships, trust, and mutual understanding. Empathy includes the ability to feel and understand the emotions of others, which can be learned and improved through various ways. Oxytocin is a neuropeptide, and its influence on social behavior and emotions has been widely studied. It is found that it can enhance emotional and cognitive empathy, as well as trust and cooperative behavior. Oxytocin acts on specific brain regions, such as the insula, amygdala, and reward circuitry, to modulate empathy-related neural processes. Oxytocin receptor gene polymorphisms are also related to empathy. Future research could explore the effect of oxytocin interventions on individuals with empathy deficiency, investigate the relationship between oxytocin receptor gene polymorphism and empathy neural networks, and study the neural mechanisms of the influence of other neurochemical substances (such as dopamine) affecting empathy. In addition, further study on empathy of typical developing individuals could provide valuable insights into the symptoms and causes of various diseases. Finally, promoting the practical application and value transformation of research results related to empathy is helpful to develop intelligent systems that can simulate human empathy and enhance human-computer interactions.

Keywords: empathy, oxytocin, social behavior, mental disorder, neurobiology

1. Introduction

Traditionally, empathy refers to the ability to share someone else's feelings or experiences by imagining what it would be like to be in that person's situation. By sharing and understanding the feelings of others, it can accurately judge the behavior of others, help obtain environmental information and adapt to the environment, and carry out social exchanges. It has important biology and sociological significance for individuals. Empathy encompasses the ability to contemplate and vicariously share in the emotional life of others and is critical for social interaction. Empathy is a multidimensional construct, which includes cognitive and affective components. Emotional empathy refers to the affective response of feeling what another person is feeling or having a similar emotional reaction. Cognitive empathy refers to the mental process of taking the perspective of another person and imagining what they are thinking or feeling. In the absence of emotional empathy, cognitive empathy is able to understand what another person is feeling without necessarily experiencing the same emotions.

Empathy is an important skill for social interactions, as it helps to build rapport, trust, and mutual understanding. Empathy can also enhance moral reasoning, ethical decision-making, and pro-social behavior. Empathy can be learned and improved through various methods, such as reading fiction, engaging in role-playing, practicing mindfulness, and receiving feedback.

The multidimensional nature of empathy has nothing to do with a single neurobiological process. Functional neuroimaging studies have shown that different components of empathy are associated with several related but distinct brain processes, marked by co-activation between brain regions [1].

1.1 The definition of empathy

Empathy is a multidimensional paradigm, and there currently is a lack of scientific consensus on its definition [2]. The debate over the concept and definition of empathy reflects the different understandings of researchers on the nature of empathy. So far, the academic understanding of the nature of empathy mainly includes the following aspects:

Some scholars consider empathy to be an instinctive response or experience of another person's emotion, an emotional phenomenon [3]. Empathy is a feeling. When the empathic person is absorbed in the empathic object, the empathic object has the expressiveness and passively produces a feeling obtained by self-imitation. Empathy is defined as the empath's emotional response to the empath's current or imminent experience of an emotion and considers an emotional response that is inconsistent with the empath's experience to be empathy, called "differential empathy". However, others define empathy as an emotional response that is considered to be similar to the emotion of the empathic object.

From the cognitive perspective, empathy is considered to be the understanding of another person's emotions from a cognitive level [4]. Empathy focuses on the understanding of another person's emotions. It is more than the sharing of emotions. Empathy is the individual's intellectual understanding of the thoughts and feelings of others, and on this basis, a cognitive empathy scale has been developed [5]. While emphasizing the cognitive component of empathy, the social role that empathy points to is more prominent. It is believed that empathy is an individual's ability to make choices about the roles of others. By understanding others' evaluations and perceptions of the outside world and predicting their further reactions from their behaviors, they can adjust their own behaviors in order to adapt quickly to society and live better in it.

Gladstein's "two-component theory" has had the most profound impact on empathy research [6]. It is believed that empathy contains both emotional and cognitive components, only that the weight of each differs in different scenarios. Gladstein divides empathy into Emotion empathy and Cognitive empathy. Emotional empathy is used to refer to an individual's ability to respond to the emotions of others. Cognitive empathy expresses an individual's ability to cognitively adapt another person's perspective and assess the state of another person. Brain imaging techniques were used to study empathy in brain-injured patients, and the results showed that emotional empathy and cognitive empathy involve separate brain regions [7]. This result provides strong evidence to support the "two-component theory".

1.2 The neural basis of empathy

There is increasing evidence suggests that empathy for pain is underpinned by neural structures [8, 9]. In the early days of research, scholars mainly explored the inherent mechanism of empathy from a cognitive level. It explains how the individual understands the emotions of others but has not explained how the individual share the emotions of others. With the advancement of medical technology, cognitive neurosciences introduce brain imaging technology into empathy research. Therefore, further deepening the research of empathy and neural mechanisms can reveal the neural basis of empathic processes and the factors that influence them.

Emotional empathy refers to the individual's experience and sharing of the emotional state of others. Studies have found that there is a "representation sharing" between individuals and others; that is, when the individual perceives the emotions or movements of others, the part that corresponds to the emotion or movement in the brain will be activated [10]. Performing sharing is considered to be the basis of empathy, which is closely related to emotional empathy [9]. In an experiment investigating olfactory empathy, researchers found that the forebrain insula region was activated when subjects smelled disgusting odors or watched videos of expressions of disgusting emotions Similar results exist in the field of nociceptive empathy research [11]. In a study of individuals receiving painful stimuli and observing others receiving painful stimuli, Singer et al. found that the anterior cingulate gyrus and anterior insula, which are responsible for emotional processing, were activated in both states in subjects [12]. The mirror neuron is a sensory-motor neuron. It is activated during both the observation and execution phases of an action [13]. Neuroscientists first identified mirror neurons in the F5 region of the premotor cortex of the rhesus monkey brain, which were activated either when the monkey performed a certain action or when they observed another similar action. In subsequent studies, researchers determined that similarly functioning brain regions exist in the human brain and called them "mirror neural systems". The discovery provides strong evidential support for representational sharing. It has also been shown that the activation of the mirror neural system is very significantly related to emotional empathy.

Representation sharing and mirror neurology explain the emotional contagion of empathy. However, emotional contagion is only the initial stage of empathy, as individuals experience similar or different emotions while being emotionally contagious to others. Emotional empathy therefore also involves the amygdala, which generates and regulates emotions. In summary, we believe that emotional empathy primarily involves the anterior insula, anterior cingulate gyrus, amygdala, and mirror neurology.

Cognitive empathy is often associated with psychological theory. Both encompass the individual's ability to understand the mental states of others and to predict their behavior accordingly, as well as to distinguish between the mental states of the self and others. Psychological theory activates mainly brain areas such as the medial prefrontal cortex, temporal pole, superior temporal sulcus, and temporoparietal junction. Schnell found that the medial prefrontal cortex, temporopolar region, temporoparietal junction, and limbic system are involved in understanding the emotional state of others [14]. The Shamay-Tsoory study on empathy in brain-injured patients showed that individuals with ventral medial prefrontal cortex injuries had poor cognitive empathy and normal emotional empathy, and those with inferior frontal gyrus injuries had poor emotional empathy and normal cognitive empathy [15].

In addition to this, the researchers found that mirror neurons, while involved in the sharing of representations, may also be involved in the understanding of the purpose and desires of others from a conscious level. Fan, Y summarized previous work and found that the dorsal anterior cingulate cortex, anterior cingulate cortex, supplementary motor areas, and bilateral insula were activated in numerous forms of empathy research [16]. Therefore, these brain regions are referred to as the core brain regions of empathy. Among them, cognitive and emotional empathy both activated the anterior insula on the left side, whereas cognitive love activated the anterior cingulate gyrus.

2. Empathy and oxytocin

Oxytocin is a nonapeptide secreted by the paraventricular and supraoptic nuclei of the hypothalamus and released in the brain and blood through the posterior lobe of the pituitary gland. It acts as a neurotransmitter and a hormone with central (amygdala, para-ventricular nucleus of the hypothalamus, hippocampus, and brain stem) and peripheral (heart, womb, and spinal cord) effects. The neuropeptide oxytocin has a solid reputation as a facilitator of social interactions. It has been shown to have various effects on our social behavior and emotions, such as increasing trust, generosity, and compassion.

Research has found that oxytocin can improve both cognitive and emotional empathy in different ways. For example, one study found that oxytocin administration increased the ability to recognize emotions in others and improved self-reported empathy [17]. Another study found that oxytocin increased the neural response to positive social feedback and reduced the neural response to negative social feedback [18]. This suggests that oxytocin can make us more sensitive to rewarding social cues and less sensitive to stressful social cues.

Studies have revealed oxytocin receptors are expressed pre- and postnatally, raising the possibility that developmental oxytocin signaling may have a lasting impact on brain organization and behavior. Oxytocin can modulate the neural activity associated with empathy in the brain. Studies have found that oxytocin can increase the activation of brain regions involved in empathy, such as the insula, the anterior cingulate cortex, and the amygdala [19]. Oxytocin can also reduce the activation of brain regions involved in self-referential processing, such as the medial prefrontal cortex, which may facilitate empathic understanding by reducing egocentric bias [20].

Oxytocin was initially thought to increase empathy for all. However, more careful research showed that although oxytocin increased empathy for in-group members, it had no affect or actually decreased empathy for out-group members [21]. A possible explanation for this behavior could be that oxytocin increases the salience of social stimuli [22]. Oxytocin does improve the ability to recognize facial emotions, but this also depends on the time and intensity of exposure [23]. More research on empathy suggests that oxytocin appears to enhance emotional empathy and has little effect on cognitive empathy [24, 25].

Research has shown, empathy is the basis for human social interaction [26]. It can help us build a trust, cooperation, empathy, and altruistic relationship. Emotional empathy can increase pro-social behavior, but this role is played with a certain degree of self-interest. When the level of personal distress in sympathizing with someone reaches a given threshold, individuals seem to gravitate toward their own feelings rather than using that distress as a motivation to help someone in need. Cognitive empathy is more conducive to pro-social behavior [27]. Pro-social behavior was marked by dorsomedial prefrontal cortex activity; this area is involved in mentalizing and helps individuals better understand the needs of others, thus enabling more altruistic behavior [27].

2.1 The impact of oxytocin level on empathy response

At present, there are two main areas of research investigating the effects of oxytocin on empathic responses. One is to examine the effect of individual endogenous oxytocin levels on the empathic response, and the other is to discuss the effect of exogenous oxytocin interventions on the empathic response.

Studies have found that the production of empathic experiences is accompanied by increased levels of oxytocin in the peripheral nervous system. By observing the empathy concern and the subjective level of individual sadness when watching short films (emotional or non-emotional), the relationship between it and plasma oxytocin level was determined. It was found that oxytocin levels in the subjects' blood were significantly higher when viewing emotional clips than nonemotional clips [28]. This suggests that the empathy experience of the individual will be accompanied by the release of the oxytocin in the brain. Interestingly, empathy-induced changes in oxytocin were associated with gender. Compared with men, the changes in oxytocin were stronger in women, and thus women's empathy is significantly higher than men's. Also, the concentration of oxytocin in an individual's blood can predict the intensity of the empathic response. During pregnancy, the mother's oxytocin level increases significantly, triggered by the rise in estrogen levels, and this increase continues into the breastfeeding period. In this process, the level of oxytocin release is strongly related to the mother's empathy [28]. Compared to mothers with low levels of oxytocin in the blood, mothers with high levels of oxytocin have stronger reward levels in the brain when they see their baby smiling. This activation suggested that they are more likely to resonate with their children's emotions [29]. In summary, there is a closely relationship between the level of endogenous oxytocin in an individual and the strength of the empathic response.

Existing neurological studies show that oxytocin ejected from the nasal cavity can directly cross the blood-brain barrier and act on the limbic systems closely related to social behavior, such as hippocampus and amygdala. By intervening with exogenous oxytocin, it is possible to directly regulate the level of oxytocin in the central nervous system and thus establish a causal relationship with society. Domes et al. were the first to report that oxytocin improved cognitive empathy ("mindreading"), as measured using the "Reading the Mind in the Eyes Test" (RMET). In the RMET, subjects were shown the eye area of different people and asked to assign one of four descriptive terms to the expression. Results found that intranasal oxytocin improved RMET performance. Another functional magnetic resonance imaging (fMRI) study investigating RMET performance depending on oxytocin Administration also detected increased insula activation (and superior temporal gyrus activation) after oxytocin administration and reproduced the oxytocin effect on RMET performance of the Domes group. Also, nasal inhalation of oxytocin directly affects the level of emotional response in empathy. However, some researchers have also found that nasal oxytocin has a significant effect on the empathy of individuals. By presenting four sad stories recording to the subjects and then recorded their empathy responses. They found that oxytocin injection increased the empathy level for the female protagonist in the story, but not for the male protagonist, regardless of the subjects' gender [30]. Thus, specifically, oxytocin can be effective in raising the level of empathic attention to the emotions of others. However, negative emotions directed toward the self due to the misfortune of others not only are not significantly enhanced but may also be weakened.

2.2 Oxytocin affects empathy response mechanisms

At present, the more recognized neural mechanism of empathy reflex is mirror neuroscience. More and more studies have found that oxytocin promotes the activity of individual mirror neurology. Oxytocin enhances the activity of mirror neurons to improve the individual's ability to imitate [31]. The fMRI was used to study the response of oxytocin in the brain of female subjects to neural network regulation of an infant crying. The results showed that injection of oxytocin weakened the activation in the right amygdala and enhanced activation of insula and inferior frontal gyrus (mirror neuron brain region). This suggests that when women perceive frightening social stimuli, oxytocin will reduce the activity of amygdala, prevent women from being distracted by anxiety or disgust, and thus promote their ability to respond to a baby's cries. At the same time, oxytocin may enhance women's empathy responses to a baby crying by enhancing the activation of brain regions (insula and inferior frontal gyrus) related to emotional empathy [32].

Oxytocin can act on the self-information processing system, weakening the self-centered tendencies of individual in social information processing and making it easier for people to shift their perspective and focus on the mental states and feelings of others [33]. A facial deformation program is used to explore the influence of oxytocin on self-others distinctions. The results of this study show that oxytocin lowers the threshold for distinguishing self from other people's faces in the experimental task and at the same time increases the perception of others' favor ability, which proves the effectiveness of oxytocin in overcoming individual self-service bias [20]. The essence of empathy is the experience of congruent emotions with the emotion-producing object. It is also found that oxytocin affects the emotional expression mechanisms of individuals, mainly by weakening negative emotions and promoting positive emotions. On the one hand, oxytocin reduces the level of amygdala activity in response to negative emotional faces (fear, anger, sadness and disgust, and physical pain) [34–37]. In addition to amygdala activity, oxytocin can also reduce the activity of other brain areas associated with negative affect, such as the anterior insula and anterior cingulate gyrus. Oxytocin may also increase activity in areas of the brain related to emotional management, such as the medial prefrontal lobe and ventrolateral prefrontal lobe. Overall, oxytocin may weaken individuals' negative emotional feelings and improve emotion management, which helps individuals reduce social anxiety and stress levels and promote positive social interactions. On the other hand, oxytocin can increase the level of activation of the reward system in social interactions. When the subjects watched positive social stimuli, such as happy faces or photos of their partners and children or participating in positive social interaction, oxytocin was found to be significantly enhanced in other brain regions of the reward system, such as the ventral segment, the nucleus accumbens, the caudate nucleus, the vomeronasal nucleus, and the midbrain [38–40]. These findings suggest that oxytocin increases the value of rewards in social contexts, enhances interpersonal connections, and strengthens social relationships.

2.3 Oxytocin regulates pro-social behavior by affecting empathy

Pro-social behavior is a common aspect of individual socialization, and it plays a vital role in individual social development. In fact, the term "pro-social behavior" was first introduced by American scholar Wispel, who contrasted it with antisocial behavior such as vandalism and aggression. According to social psychologists, pro-social behavior aligns with social moral standards, whereas antisocial behavior violates them. Initially, pro-social behavior refers to adherence to social rules and moral standards. It involves voluntary actions that individuals take to benefit others, the community, or society as a whole, including acts of cooperation, assistance, sharing, donation, and compassion, among others. Such behaviors are subject to the constraints of social moral standards. There are two types of pro-social behavior, categorized according to the situation. Pro-social behavior in emergency situations involves acts of courage, such as saving lives during water or fire emergencies, while pro-social behavior in non-emergency situations entails helping others without harm or threat to the person performing the behavior. Pro-social behavior can be motivated by two different factors. Completely pro-social behavior occurs when individuals help others without any conscious pursuit of personal benefit. Conversely, partial pro-social behavior involves individuals helping others with the expectation of receiving help in return. The social heuristic hypothesis suggests that pro-social behaviors become internalized as default heuristics in an individual's daily life, leading to the development of automatic pro-social behavioral tendencies. Ultimately, pro-social behavior is an essential aspect of individual socialization and development, and it plays a crucial role in maintaining social cohesion and stability.

Batson proposed the empathy-altruism hypothesis, which suggests that pro-social behavior motivated by empathy aims to provide welfare to those in need [41]. Several scholars have provided evidence that empathy is the biological foundation of pro-social behavior, altruism, and morality. The empathy-altruism hypothesis predicts that empathically aroused individuals will experience empathic joy when they learn that the needs of others have been met, but this joy is a consequence of their helping to address the needs of others and is not an end in itself. Empathy not only has an altruistic effect; it can also have an egoistic effect. Empathic emotions can be unpleasant, and individuals can reduce the experience of discomfort by engaging in pro-social behavior. Studies related to pro-social behavior have demonstrated that empathy is a vital factor that influences pro-social behavior. Empathy enables people to understand the feelings and needs of others, which can motivate individuals to exhibit more pro-social behaviors.

Oxytocin is believed to be closely related to empathy and is also one of the most extensively studied hormones. Pregnant women have the effect of promoting uterine contractions during childbirth and lactation after childbirth. In addition, the establishment of mother-child attachment relationships related to empathy, children playing together, making friends, spouses, and erections during sexual activity all require oxytocin. Therefore, some scholars suggest calling it "empathy". Because oxytocin can be sprayed through the nasal mucosa and enter the cerebrospinal fluid through the olfactory filament to act on sympathetic brain regions, it can also promote the release of oxytocin from oxytocin neurons in the paraventricular nucleus and supraoptic nucleus of the hypothalamus into the circulating blood through the posterior pituitary, playing a "social" hormone role. Literature shows that nasal spraying of oxytocin can enhance social skills such as pain empathy, mother-child attachment, social cognition, and facial recognition, enhancing empathy within groups, integrity, collaboration, conformity, and resistance to attacks outside of groups.

2.3.1 Oxytocin and face perception

Numerous studies have demonstrated that oxytocin can impact face recognition abilities. Specifically, oxytocin has been found to have a significant effect on individuals' ability to recognize basic emotions. In a study by Lischke et al., participants from both the oxytocin and placebo groups were given dynamic facial expression recognition tasks involving four emotions—joy, anger, sadness, and fear—with varying levels of emotional intensity [42]. The results indicated that the oxytocin group had a more accurate recognition of all emotions, especially recognizing lower intensity facial expressions, compared to the placebo group.

Additionally, some clinical studies have suggested that oxytocin might improve emotional recognition ability in patients with certain mental disorders. For instance, exogenous oxytocin attenuates early attentional biases toward negative stimuli and increases selective attention and recognition of emotional cues in faces, particularly around the eyes [43]. However, some studies have shown that the impact of oxytocin on emotional recognition is regulated by the emotional valence, meaning that it enhances recognition of positive emotions while reducing recognition of negative emotions. For instance, Domes instructed male participants to focus on emotional faces and found that oxytocin increased the frequency of gaze on positive faces (happy) but decreased the frequency on negative faces (angry) [44]. Woolley et al. found that oxytocin can improve facial recognition ability but cannot improve interpersonal trust [45]. While oxytocin appears to have a positive effect on the recognition of certain emotions, further research is needed to confirm these findings.

2.3.2 Oxytocin and emotional inference

Emotional inference refers to people's ability to understand and perceive the emotional state of others in social interactions. When we communicate with others, we usually infer their emotional state through nonverbal signals such as facial expressions, language intonation, and body language, such as joy, sadness, and anger. This ability is crucial for establishing and maintaining social interaction and interpersonal relationships. Emotional inference includes not only the perception of the emotional state of others but also the understanding and inference of their emotional state. For example, when we see a person smiling, we can not only perceive that their emotional state is positive but also infer that they may be smiling due to reasons such as happiness, satisfaction, and gratitude. Emotional inference is an advanced cognitive function, which involves the synergy of multiple brain regions, including the visual cortex, amygdala, prefrontal cortex, and so on. In daily life, emotional inference ability significantly impacts people's social and emotional health. Some social and emotional disorders, such as autism and depression, are related to insufficient emotional inference ability. Therefore, improving emotional inference ability is of great significance for improving social and emotional disorders.

Observing basic emotional information such as facial expressions, speech, and actions can trigger an individual's emotional experiences due to a process called emotional resonance. The Perception-Action Model (PAM) of empathy proposes that when someone perceives sensory emotional information, they unconsciously engage in an action imitation process that activates the mirror neuron system. This system includes brain areas such as the parietal and inferior lobules, inferior frontal gyrus, and anterior motor cortex, which enable the perceiver to feel the perceived emotion through physiological feedback [46].

Similarly, Walter's Empathy Loop Model suggests that individuals must infer and understand others' psychological states to resonate emotionally with them [47]. This process activates the ventromedial prefrontal lobe and enables emotional resonance with the other person. Thus, individuals who are better at taking the perspective of others are more capable of projecting themselves into others' psychological states and sharing their emotions. Those who struggle to overcome their egocentric tendency may find it challenging to empathize with others [48]. Overall, emotional resonance is a complex and embodied process that involves both physiological and cognitive mechanisms.

Empathic reactions can lead to the development of empathy concerns toward others, resulting in positive emotions such as love, care, and other motivational factors that promote pro-social behavior. This positive attention to others' misfortunes is an essential psychological factor that fosters empathy and pro-social behavior.

Researchers have found that oxytocin can enhance an individual's ability to infer others' emotions, as demonstrated through experimental paradigms such as the RMET and viewpoint-taking tasks. For instance, Domes discovered that oxytocin administration through nasal injection improved male participants' theory-of-mind abilities in the RMET [49]. The task requires participants to infer the psychological state of individuals based solely on their eye area in a photograph, and oxytocin improved accuracy, particularly in more challenging tasks [24].

In summary, oxytocin has a positive impact on emotional inference. This discovery helps us better understand the role of oxytocin in social behavior and emotions and provides a more in-depth research direction for future research.

2.3.3 Oxytocin and trusting and cooperative behavior

Oxytocin also has the ability to influence trusting and cooperative behaviors, thereby facilitating social interactions and building strong interpersonal relationships. The neural mechanisms underlying cooperative and protective behavior, as an important function of human survival, have always been a focus of discussion among researchers. Related research has found that compared to making non cooperative choices, when individuals make cooperative choices that require loss of their own interests but result in greater benefits for the team, oxytocin group participants tend to choose more cooperative options [50, 51], which means that oxytocin makes individuals more willing to sacrifice their own interests to promote cooperative behavior. More options have been chosen for noncooperation to ensure that members of the inner group are not attacked.

The Trust Game is a widely used research paradigm for investigating human trust behavior, with investment amount and frequency serving as key indicators. In a pioneering study, Kosfeld et al. administered nasal spray oxytocin and found that it increased investment behavior and trust tendencies in individuals, demonstrating the influence of oxytocin on trust behavior [52]. However, subsequent studies indicate that the trustworthiness of the trustee can affect the effectiveness of oxytocin, with oxytocin enhancing trust behavior only toward trustworthy trustees [53]. Moreover, oxytocin seems to affect men and women differently, as it causes women to exhibit less trust and forgiveness toward betraying trustees than men [54]. Additionally, personality traits play a crucial role in determining trust behavior, with oxytocin causing individuals with borderline personality disorder (BDP) to exhibit less trust [55].

Apart from trust games, researchers have explored the impact of oxytocin on trust behavior in other paradigms. In one study, participants evaluated strangers' photos to measure trust levels, and the oxytocin group displayed higher trust scores and was rated as more attractive than the placebo group [56]. Moreover, pupil contraction in trustees can affect an individual's trust level, with oxytocin reducing trust in pupillary contraction partners but increasing trust in pupillary dilation partners [57]. Long-term (two-week) intranasal administration of oxytocin has also been found to reduce avoidant attachment and increase attachment to peers, making participants more inclined to trust others [58]. Neuroimaging studies provide further evidence for the mechanism by which oxytocin impacts trust behavior. In functional magnetic resonance imaging (fMRI) research, oxytocin reduced the activation of the striatum, amygdala, and midbrain regions after betrayal, indicating less fear processing and behavior adjustment based on feedback information and thus more trust behavior [59]. Additionally, oxytocin promotes mutually beneficial cooperative behavior by increasing the value of individuals' rewards and forming the belief that others are trustworthy, as observed in stronger caudate nucleus activation in the oxytocin group during cooperation [60].

While many studies have shown that oxytocin enhances trust behavior, replicating these results can be challenging due to factors such as variation in experimental design and procedures. A meta-analysis did not find any significant effect of nasal administration of oxytocin on trust behavior [61]. This suggests that further research is needed to replicate and validate the experimental effects of oxytocin on trust behavior in the future.

2.3.4 Oxytocin and group preference

Group preference, also known as herding behavior or social assimilation, is the tendency for people to accept the ideas, values, and behaviors of a group when they are in that group, rather than pursuing their own unique ideas and behaviors. This phenomenon occurs in a variety of different groups, such as families, schools, workplaces, social media, and so forth. It usually occurs when people want to gain social acceptance and group approval or avoid social exclusion. When people feel different or isolated from others, they usually try to adjust their behaviors and thoughts to conform to the group's expectations and standards. Current research suggests that oxytocin can influence group preferences, making it easier for people to accept and follow group expectations and behavioral norms.

Variations of the Prisoner's Dilemma task have been used by researchers to investigate cooperative and protective behavior, and the results indicate that the oxytocin group exhibited greater love for inner group members and lower selfishness but did not show any impact on their hatred toward outer group members [62]. Interestingly, oxytocin did not display the same tendency as aggression-enhancing hormones toward members outside the group. In the Prisoner's Dilemma task, De Dreu found that individuals in the oxytocin group were more likely to choose not to cooperate with external group members to ensure their higher interests, resulting in lower profits for internal group members in a "predatory" state [62]. Furthermore, in situations of "greed," where individuals choose not to cooperate to improve their own interests and reduce the profits of members outside the group, the oxytocin group did not increase the number of noncooperative choices to demonstrate "plundering" of members outside the group. This suggests that oxytocin promotes the protection of more vulnerable internal group members, rather than attacking external group members. This further supports the idea that oxytocin promotes love for members within the group, rather than hatred for members outside the group.

Similarly, additional studies that added cognitive load to the Prisoner's Dilemma task found that oxytocin promotes cooperative behavior without the influence of cognitive load. The placebo group, on the other hand, exhibited more cooperative behavior only when there was cognitive load. These results suggest that the love for members of the inner group induced by oxytocin is inherent and intuitive, rather than intentional [63].

However, subsequent studies have found a "dark side" to oxytocin, such as an increase in distrustful [64] and deceptive [65] decision-making behaviors, suggesting that oxytocin does not always promote pro-social behavior and that the factors influencing it have attracted the attention of researchers.

2.4 Oxytocin, social cognition, and neural circuits patterns

The effects of oxytocin on individual social cognition may be achieved mainly through modulating the activation patterns of the amygdala and reward system [66, 67]. Specifically, oxytocin weakens the amygdala response to negative emotional information and increases its functional connectivity to brain regions involved in emotion management. Experimental studies have found that the administration of intranasal oxytocin reduces amygdala activation after subjects are confronted with fearful faces and disgusting scenes. Similarly, in the fMRI paradigm study, the amygdala showed an increased responsiveness to the fearful whites of the eyes. However, this response was diminished when oxytocin was given prior to testing [35]. Interestingly, when different stimuli are presented, oxytocin appears to not only reduce amygdala activation but also facilitate insula responses, thus making negative stimuli easier to remember [68].

Oxytocin is closely associated with the reward neural circuit, which is concentrated in the midbrain, including the ventral tegmental area (VTA) and the basal ganglia, which contains the globus pallidus (GP), the substantia nigra (SN), the ventral striatum (VS), and the dorsal striatum, which contains the putamen and caudate, areas that are essential for reward processing. When cooperative behavior in a Prisoner Dilemma game was reciprocated, oxytocin augmented caudate nucleus activation in addition to the amygdala response. Oxytocin acts in these brain areas above to increase reward and emergence in interpersonal interactions (face processing, cooperation, romance, parenting, etc.), thus promoting social bonding and attachment.

2.5 Oxytocin receptor and empathy disorder

Oxytocin receptor is widely expressed in mammalian brains, including the septum, nucleus accumbens, and ventral tegmental area. Research finds a correlation between the oxytocin receptor gene and psychopathy traits, suggesting that the oxytocin gene can affect the brain's empathy response [69].

The oxytocin receptor gene, which is located on chromosome 3p25.3, has been implicated as a candidate gene for susceptibility of autism spectrum disorder (ASD) [70]. ASD is a representative disorder of pervasive developmental disorders, which affect the development of social, communication, and behavioral skills. Its core symptoms are: impaired social interaction, communication difficulties, narrow interests, and stereotypical repetitive behavior patterns. Psychologists generally agree that autism is the result of a combination of genetics and environment. However, in recent years, the emerging empathic-systemic theory has explained the onset of autism in a more comprehensive way. Children with autism have difficulty recognizing facial expressions, have a lower ability to recognize the emotions of others, and have a lower level of development in naming and matching expressions than normal children. This can lead to difficulties in accurately perceiving the emotions of others or biases in their perception of the emotions of others, resulting in deficits in empathy.

Experiments have shown that children with autism have lower plasma oxytocin levels than healthy children of the same age and that elevated oxytocin levels do not correlate positively with increasing age of the child. Lower levels of endogenous oxytocin are arousing in children with ASD, but not in adolescents or adults [71]. Anyway, oxytocin is generally considered to be associated with social deficits in ASD.

Similarly, animal experiments have shown that oxytocin receptor knockout mice exhibit behavioral deficits associated with autism and that exogenous oxytocin supplementation ameliorates this deficit [72, 73]. Autism is associated with mutations in genes such as the gene encoding contact-related protein-like protein 2 (Cntnap2). Intraperitoneal or intranasal administration of oxytocin ameliorates social behavioral deficits in mice with null mutations in the Cntnap2 gene. The use of drugs to promote the release of endogenous oxytocin can stimulate the oxytocin system more effectively and improve socially abnormal behavior.

3. Discussion and outlook

Empathy is a universal emotion that has evolved to facilitate pro-social behavior and interpersonal interactions in humans. Empathy can be divided into emotional empathy and cognitive empathy. Emotional empathy refers to an individual's emotion infection and recognition of emotions, which emerges in infancy. The main subcortical brain areas involved are: the insula (Nsula), the anterior cingulate gyrus (ACC), and the mirror neurology (MNS). Cognitive empathy refers to an individual's understanding of the emotions of others and involves the main brain region: the ventral medial prefrontal lobe (vm PFC). The neural network of empathy develops over time with age and matures in early adulthood. The neural network of empathy is modulated by cognitive appraisal. Oxytocin enhances empathy. Oxytocin receptor gene polymorphisms are associated with empathy. Oxytocin may facilitate empathic responses by enhancing insula and subfrontal gyrus activation. Further research could be conducted in the future in the following areas:

3.1 Conducting an intervention study of oxytocin to improve empathy-deficient individuals

Hurlemann demonstrated that oxytocin primarily promotes affective empathy without affecting cognitive empathy. In contrast, Pedersen found that oxytocin improved the theory of mind levels (including cognitive empathy) in patients with schizophrenia, demonstrating that oxytocin can also promote cognitive empathy. The reasons for the inconsistency between these two findings have not yet been explored in relevant studies. Therefore, the different effects of oxytocin on emotional and cognitive empathy could be investigated in the future.

Oxytocin has been found to improve emotion recognition in adolescent males with autism [74] and to improve empathic accuracy in individuals with poor social competence [75]. The study of the ameliorative effects of oxytocin on empathy-deficient individuals (autism) has facilitated the development of interventions for empathy-deficient individuals. Therefore, more research should be conducted in the future on the effects of oxytocin interventions on empathy-deficient individuals. In addition, most studies on the effects of oxytocin on empathy have used male subjects. In the future, the mechanism of oxytocin on empathy could be investigated using female subjects or a mixture of male and female subjects.

3.2 Combining brain imaging and genetic techniques to study the relationship between genetic polymorphisms and empathic neural networks

Studies have found that oxytocin receptor gene polymorphisms are associated with empathy [76, 77]. However, the relationship between gene polymorphisms and empathic neural networks is unclear.

Walter et al. were the first to use brain imaging genetic techniques to study the relationship between genetic polymorphisms and empathic neural networks in psychiatric patients and found dysfunctional cognitive empathic neural networks in psychiatric patients carrying certain genotypes. However, there are currently few such studies. Therefore, in the future, brain imaging and genetic techniques can be combined to study the relationship between genetic polymorphisms and empathic neural networks and thus discover the relationship between genes, neural networks and empathy. In addition, Lackner et al. found that dopamine was associated with the level of theory of mind in preschool children, thus promoting cognitive empathy [78]. Dopamine plays an important role in the maturation of the prefrontal lobe of the brain, which is the main brain region involved in cognitive empathy. Therefore, the neural mechanisms underlying the effects of other neurochemicals on empathy, such as dopamine, could be investigated in the future.

3.3 In-depth understanding of empathy in atypically developing individuals

Currently, many researchers are focusing on atypical developmental individuals with autism spectrum disorders, attention deficit hyperactivity disorder, and schizophrenia spectrum disorders and how their empathic abilities differ from those of typically developing individuals. They aim to identify the brain regions that are specifically involved in emotional or cognitive empathy and how this knowledge can help them understand the symptoms and etiology of these disorders, thereby providing better support for individuals with atypical development. Therefore, this direction is likely to remain an important element of empathy research in the future.

3.4 Promoting the practical application and value translation of research findings related to the neural basis of empathy

The exploration of the different components of empathy or the cognitive neural bases underlying empathic processing can, on the one hand, contribute to a deeper knowledge and understanding of the theories related to this field. It also facilitates the application of the theoretical framework related to empathy to the needs of the times and society in social practice. For example, an understanding of traditional empathy theoretical models and their neural underpinnings can help in the development and design of intelligences in an era when intelligences are becoming more and more relevant to people's lives. Through the establishment of an intelligent body empathy and its computational model, simulating the expression of emotions and the recognition and understanding of emotions involved in human-computer interaction situations, the intelligent body becomes a more ideal "companion" or "assistant" in human life.

Author details

Guangxin Yue
Institute of Basic Research in Clinical Medicine, China Academy of Chinese Medical Sciences, Beijing, China

*Address all correspondence to: yuegx73@hotmail.com

References

[1] Lamm C, Rütgen M, Wagner IC. Imaging empathy and prosocial emotions. Neuroscience Letters. 2019;**693**:49-53

[2] Uysal N, Çamsari UM, Ateş M, Kandİş S, Karakiliç A, Çamsari GB. Empathy as a concept from bench to bedside: A translational challenge. Noro Psikiyatri Arsivi. 2020;**57**(1):71-77

[3] Wondra JD, Ellsworth PC. An appraisal theory of empathy and other vicarious emotional experiences. Psychological Review. 2015;**122**(3):411-428

[4] Goodhew SC, Edwards M. Attentional control both helps and harms empathy. Cognition. 2021;**206**:104505

[5] Liu J, Qiao X, Dong F, Raine A. The Chinese version of the cognitive, affective, and somatic empathy scale for children: Validation, gender invariance and associated factors. PLoS One. 2018;**13**(5):e0195268

[6] Erol A, Akyalcin Kirdok A, Zorlu N, Polat S, Mete L. Empathy, and its relationship with cognitive and emotional functions in alcohol dependency. Nordic Journal of Psychiatry. 2017;**71**(3):205-209

[7] Driscoll DM, Dal Monte O, Solomon J, Krueger F, Grafman J. Empathic deficits in combat veterans with traumatic brain injury: A voxel-based lesion-symptom mapping study. Cognitive and Behavioral Neurology: Official Journal of the Society for Behavioral and Cognitive Neurology. 2012;**25**(4):160-166

[8] Lamm C, Decety J, Singer T. Meta-analytic evidence for common and distinct neural networks associated with directly experienced pain and empathy for pain. NeuroImage. 2011;**54**(3):2492-2502

[9] Betti V, Aglioti SM. Dynamic construction of the neural networks underpinning empathy for pain. Neuroscience and Biobehavioral Reviews. 2016;**63**:191-206

[10] Hartmann H, Riva F, Rütgen M, Lamm C. Placebo analgesia does not reduce empathy for naturalistic depictions of Others' pain in a somatosensory specific way. Cerebral Cortex Communications. 2021;**2**(3):tgab039

[11] Cheng Y, Lin CP, Liu HL, Hsu YY, Lim KE, Hung D, et al. Expertise modulates the perception of pain in others. Current Biology: CB. 2007;**17**(19):1708-1713

[12] Ong WY, Stohler CS, Herr DR. Role of the prefrontal cortex in pain processing. Molecular Neurobiology. 2019;**56**(2):1137-1166

[13] Bekkali S, Youssef GJ, Donaldson PH, Albein-Urios N, Hyde C, Enticott PG. Is the putative Mirror neuron system associated with empathy? A systematic review and meta-analysis. Neuropsychology Review. 2021;**31**(1):14-57

[14] Schnell K, Bluschke S, Konradt B, Walter H. Functional relations of empathy and mentalizing: An fMRI study on the neural basis of cognitive empathy. NeuroImage. 2011;**54**(2):1743-1754

[15] Shamay-Tsoory SG, Aharon-Peretz J, Perry D. Two systems for empathy: A double dissociation between emotional

and cognitive empathy in inferior frontal gyrus versus ventromedial prefrontal lesions. Brain: A Journal of Neurology. 2009;**132**(Pt 3):617-627

[16] Fan Y, Han S. Temporal dynamic of neural mechanisms involved in empathy for pain: An event-related brain potential study. Neuropsychologia. 2008;**46**(1):160-173

[17] Leppanen J, Ng KW, Tchanturia K, Treasure J. Meta-analysis of the effects of intranasal oxytocin on interpretation and expression of emotions. Neuroscience and Biobehavioral Reviews. 2017;**78**:125-144

[18] Aupperle RL, Morris AS, Silk JS, Criss MM, Judah MR, Eagleton SG, et al. Neural responses to maternal praise and criticism: Relationship to depression and anxiety symptoms in high-risk adolescent girls. NeuroImage Clinical. 2016;**11**:548-554

[19] Yamagishi A, Lee J, Sato N. Oxytocin in the anterior cingulate cortex is involved in helping behaviour. Behavioural Brain Research. 2020;**393**:112790

[20] Zhao W, Yao S, Li Q, Geng Y, Ma X, Luo L, et al. Oxytocin blurs the self-other distinction during trait judgments and reduces medial prefrontal cortex responses. Human Brain Mapping. 2016;**37**(7):2512-2527

[21] De Dreu CK, Kret ME. Oxytocin conditions intergroup relations through upregulated In-group empathy, cooperation, conformity, and defense. Biological Psychiatry. 2016;**79**(3):165-173

[22] Ne'eman R, Perach-Barzilay N, Fischer-Shofty M, Atias A, Shamay-Tsoory SG. Intranasal administration of oxytocin increases human aggressive behavior. Hormones and Behavior. 2016;**80**:125-131

[23] Shahrestani S, Kemp AH, Guastella AJ. The impact of a single administration of intranasal oxytocin on the recognition of basic emotions in humans: A meta-analysis. Neuropsychopharmacology: Official Publication of the American College of Neuropsychopharmacology. 2013;**38**(10):1929-1936

[24] Domes G, Ower N, von Dawans B, Spengler FB, Dziobek I, Bohus M, et al. Effects of intranasal oxytocin administration on empathy and approach motivation in women with borderline personality disorder: A randomized controlled trial. Translational Psychiatry. 2019;**9**(1):328

[25] Geng Y, Zhao W, Zhou F, Ma X, Yao S, Hurlemann R, et al. Oxytocin enhancement of emotional empathy: Generalization across cultures and effects on amygdala activity. Frontiers in Neuroscience. 2018;**12**:512

[26] Williams A, O'Driscoll K, Moore C. The influence of empathic concern on prosocial behavior in children. Frontiers in Psychology. 2014;**5**:425

[27] Waytz A, Zaki J, Mitchell JP. Response of dorsomedial prefrontal cortex predicts altruistic behavior. The Journal of Neuroscience: The Official Journal of the Society for Neuroscience. 2012;**32**(22):7646-7650

[28] Barraza JA, Zak PJ. Empathy toward strangers triggers oxytocin release and subsequent generosity. Annals of the New York Academy of Sciences. 2009;**1167**:182-189

[29] Strathearn L, Fonagy P, Amico J, Montague PR. Adult attachment predicts maternal brain and oxytocin response to

infant cues. Neuropsychopharmacology: Official Publication of the American College of Neuropsychopharmacology. 2009;**34**(13):2655-2666

[30] Palgi S, Klein E, Shamay-Tsoory SG. Intranasal administration of oxytocin increases compassion toward women. Social Cognitive and Affective Neuroscience. 2015;**10**(3):311-317

[31] Perry A, Bentin S, Shalev I, Israel S, Uzefovsky F, Bar-On D, et al. Intranasal oxytocin modulates EEG mu/alpha and beta rhythms during perception of biological motion. Psychoneuroendocrinology. 2010;**35**(10):1446-1453

[32] Riem MM, Bakermans-Kranenburg MJ, Pieper S, Tops M, Boksem MA, Vermeiren RR, et al. Oxytocin modulates amygdala, insula, and inferior frontal gyrus responses to infant crying: A randomized controlled trial. Biological Psychiatry. 2011;**70**(3):291-297

[33] Colonnello V, Chen FS, Panksepp J, Heinrichs M. Oxytocin sharpens self-other perceptual boundary. Psychoneuroendocrinology. 2013;**38**(12):2996-3002

[34] Sauer C, Montag C, Reuter M, Kirsch P. Imaging oxytocin × dopamine interactions: An epistasis effect of CD38 and COMT gene variants influences the impact of oxytocin on amygdala activation to social stimuli. Frontiers in Neuroscience. 2013;7:45

[35] Kanat M, Heinrichs M, Mader I, van Elst LT, Domes G. Oxytocin modulates amygdala reactivity to masked fearful eyes. Neuropsychopharmacology: Official Publication of the American College of Neuropsychopharmacology. 2015;**40**(11):2632-2638

[36] Labuschagne I, Phan KL, Wood A, Angstadt M, Chua P, Heinrichs M, et al. Medial frontal hyperactivity to sad faces in generalized social anxiety disorder and modulation by oxytocin. The International Journal of Neuropsychopharmacology. 2012;**15**(7):883-896

[37] Zunhammer M, Geis S, Busch V, Greenlee MW, Eichhammer P. Effects of intranasal oxytocin on thermal pain in healthy men: A randomized functional magnetic resonance imaging study. Psychosomatic Medicine. 2015;**77**(2):156-166

[38] Gamer M, Zurowski B, Büchel C. Different amygdala subregions mediate valence-related and attentional effects of oxytocin in humans. Proceedings of the National Academy of Sciences of the United States of America. 2010;**107**(20):9400-9405

[39] Scheele D, Wille A, Kendrick KM, Stoffel-Wagner B, Becker B, Güntürkün O, et al. Oxytocin enhances brain reward system responses in men viewing the face of their female partner. Proceedings of the National Academy of Sciences of the United States of America. 2013;**110**(50):20308-20313

[40] Wittfoth-Schardt D, Gründing J, Wittfoth M, Lanfermann H, Heinrichs M, Domes G, et al. Oxytocin modulates neural reactivity to children's faces as a function of social salience. Neuropsychopharmacology: Official Publication of the American College of Neuropsychopharmacology. 2012;**37**(8):1799-1807

[41] Batson CD, Batson JG, Slingsby JK, Harrell KL, Peekna HM, Todd RM. Empathic joy and the empathy-altruism hypothesis. Journal of Personality and Social Psychology. 1991;**61**(3):413-426

[42] Lischke A, Berger C, Prehn K, Heinrichs M, Herpertz SC, Domes G. Intranasal oxytocin enhances emotion recognition from dynamic facial expressions and leaves eye-gaze unaffected. Psychoneuroendocrinology. 2012;**37**(4):475-481

[43] Ellenbogen MA. Oxytocin and facial emotion recognition. Current Topics in Behavioral Neurosciences. 2018;**35**:349-374

[44] Domes G, Steiner A, Porges SW, Heinrichs M. Oxytocin differentially modulates eye gaze to naturalistic social signals of happiness and anger. Psychoneuroendocrinology. 2013;**38**(7):1198-1202

[45] Woolley JD, Chuang B, Fussell C, Scherer S, Biagianti B, Fulford D, et al. Intranasal oxytocin increases facial expressivity, but not ratings of trustworthiness, in patients with schizophrenia and healthy controls. Psychological Medicine. 2017;**47**(7):1311-1322

[46] Molenberghs P, Cunnington R, Mattingley JB. Brain regions with mirror properties: A meta-analysis of 125 human fMRI studies. Neuroscience and Biobehavioral Reviews. 2012;**36**(1):341-349

[47] Walter H. Social cognitive neuroscience of empathy: Concepts, circuits, and genes. Emotion Review. 2012;**4**(1):9-17

[48] Yue T, Pan W, Huang X. The relationship between trait positive empathy and brain structure: A voxel-based morphometry study. Neuroreport. 2016;**27**(6):422-426

[49] Domes G, Kumbier E, Heinrichs M, Herpertz SC. Oxytocin promotes facial emotion recognition and amygdala reactivity in adults with asperger syndrome. Neuropsychopharmacology: Official Publication of the American College of Neuropsychopharmacology. 2014;**39**(3):698-706

[50] Declerck CH, Boone C, Kiyonari T. The effect of oxytocin on cooperation in a prisoner's dilemma depends on the social context and a person's social value orientation. Social Cognitive and Affective Neuroscience. 2014;**9**(6):802-809

[51] De Wilde TR, Ten Velden FS, De Dreu CK. The neuropeptide oxytocin enhances information sharing and group decision making quality. Scientific Reports. 2017;7:40622

[52] Kosfeld M, Heinrichs M, Zak PJ, Fischbacher U, Fehr E. Oxytocin increases trust in humans. Nature. 2005;**435**(7042):673-676

[53] Mikolajczak M, Gross JJ, Lane A, Corneille O, de Timary P, Luminet O. Oxytocin makes people trusting, not gullible. Psychological Science. 2010;**21**(8):1072-1074

[54] Yao S, Zhao W, Cheng R, Geng Y, Luo L, Kendrick KM. Oxytocin makes females, but not males, less forgiving following betrayal of trust. The International Journal of Neuropsychopharmacology. 2014;**17**(11):1785-1792

[55] Ebert A, Kolb M, Heller J, Edel MA, Roser P, Brüne M. Modulation of interpersonal trust in borderline personality disorder by intranasal oxytocin and childhood trauma. Social Neuroscience. 2013;**8**(4):305-313

[56] Theodoridou A, Rowe AC, Penton-Voak IS, Rogers PJ. Oxytocin and social perception: Oxytocin increases

perceived facial trustworthiness and attractiveness. Hormones and Behavior. 2009;**56**(1):128-132

[57] Kret ME, De Dreu CK. Pupil-mimicry conditions trust in partners: Moderation by oxytocin and group membership. Proceedings Biological Sciences. 2017;**284**(1850):20162554

[58] Bernhard RM, Chaponis J, Siburian R, Gallagher P, Ransohoff K, Wikler D, et al. Variation in the oxytocin receptor gene (OXTR) is associated with differences in moral judgment. Social Cognitive and Affective Neuroscience. 2016;**11**(12):1872-1881

[59] Baumgartner T, Heinrichs M, Vonlanthen A, Fischbacher U, Fehr E. Oxytocin shapes the neural circuitry of trust and trust adaptation in humans. Neuron. 2008;**58**(4):639-650

[60] Rilling JK, DeMarco AC, Hackett PD, Thompson R, Ditzen B, Patel R, et al. Effects of intranasal oxytocin and vasopressin on cooperative behavior and associated brain activity in men. Psychoneuroendocrinology. 2012;**37**(4):447-461

[61] Nave G, Camerer C, McCullough M. Does oxytocin increase Trust in Humans? A critical review of research. Perspectives on Psychological Science: A Journal of the Association for Psychological Science. 2015;**10**(6):772-789

[62] De Dreu CK, Greer LL, Handgraaf MJ, Shalvi S, Van Kleef GA, Baas M, et al. The neuropeptide oxytocin regulates parochial altruism in intergroup conflict among humans. Science (New York, N.Y.). 2010;**328**(5984):1408-1411

[63] Daughters K, Manstead ASR, Ten Velden FS, De Dreu CKW. Oxytocin modulates third-party sanctioning of selfish and generous behavior within and between groups. Psychoneuroendocrinology. 2017;**77**:18-24

[64] De Dreu CK, Shalvi S, Greer LL, Van Kleef GA, Handgraaf MJ. Oxytocin motivates non-cooperation in intergroup conflict to protect vulnerable in-group members. PLoS One. 2012;**7**(11):e46751

[65] Shalvi S, De Dreu CK. Oxytocin promotes group-serving dishonesty. Proceedings of the National Academy of Sciences of the United States of America. 2014;**111**(15):5503-5507

[66] Adolphs R. What does the amygdala contribute to social cognition? Annals of the New York Academy of Sciences. 2010;**1191**(1):42-61

[67] Musardo S, Contestabile A, Knoop M, Baud O, Bellone C. Oxytocin neurons mediate the effect of social isolation via the VTA circuits. eLife. 2022;**11**:e73421

[68] Striepens N, Scheele D, Kendrick KM, Becker B, Schäfer L, Schwalba K, et al. Oxytocin facilitates protective responses to aversive social stimuli in males. Proceedings of the National Academy of Sciences of the United States of America. 2012;**109**(44):18144-18149

[69] Dadds MR, Moul C, Cauchi A, Dobson-Stone C, Hawes DJ, Brennan J, et al. Polymorphisms in the oxytocin receptor gene are associated with the development of psychopathy. Development and Psychopathology. 2014;**26**(1):21-31

[70] Liu X, Kawamura Y, Shimada T, Otowa T, Koishi S, Sugiyama T, et al. Association of the oxytocin receptor (OXTR) gene polymorphisms with autism spectrum disorder (ASD) in the Japanese population. Journal of Human Genetics. 2010;**55**(3):137-141

[71] Green L, Fein D, Modahl C, Feinstein C, Waterhouse L, Morris M. Oxytocin and autistic disorder: Alterations in peptide forms. Biological Psychiatry. 2001;**50**(8):609-613

[72] Latt HM, Matsushita H, Morino M, Koga Y, Michiue H, Nishiki T, et al. Oxytocin inhibits corticosterone-induced apoptosis in primary hippocampal neurons. Neuroscience. 2018;**379**:383-389

[73] Pati D, Harden SW, Sheng W, Kelly KB, de Kloet AD, Krause EG, et al. Endogenous oxytocin inhibits hypothalamic corticotrophin-releasing hormone neurones following acute hypernatraemia. Journal of Neuroendocrinology. 2020;**32**(3):e12839

[74] Guastella AJ, Einfeld SL, Gray KM, Rinehart NJ, Tonge BJ, Lambert TJ, et al. Intranasal oxytocin improves emotion recognition for youth with autism spectrum disorders. Biological Psychiatry. 2010;**67**(7):692-694

[75] Bartz JA, Nitschke JP, Krol SA, Tellier PP. Oxytocin selectively improves empathic accuracy: A replication in men and novel insights in women. Biological Psychiatry Cognitive Neuroscience and Neuroimaging. 2019;**4**(12):1042-1048

[76] Rodrigues SM, Saslow LR, Garcia N, John OP, Keltner D. Oxytocin receptor genetic variation relates to empathy and stress reactivity in humans. Proceedings of the National Academy of Sciences of the United States of America. 2009;**106**(50):21437-21441

[77] Wu N, Li Z, Su Y. The association between oxytocin receptor gene polymorphism (OXTR) and trait empathy. Journal of Affective Disorders. 2012;**138**(3):468-472

[78] Lackner CL, Bowman LC, Sabbagh MA. Dopaminergic functioning and preschoolers' theory of mind. Neuropsychologia. 2010;**48**(6):1767-1774

Chapter 5

OXTR Gene Polymorphisms and Event-Related Potentials in Humans: A Systematic Review

Diego Armando León-Rodríguez, Julian Becerra, Juan Carlos Caicedo Mera, Luis Fernando Cardenas, Jorge Martínez Cotrina and Diego Mauricio Aponte Canencio

Abstract

Oxytocin receptor (OXTR) gene polymorphisms have been consistently associated with humans' differences in sensitivity to social cues, social cognition, stress response, and brain activity. However, how social and affective neural processing differs across carriers of distinct OXTR gene polymorphisms remains unclear. This systematic PRISMA review is the first to examine the experimental literature on the relationship between OXTR polymorphisms and ERP components. Eight studies published between 2014 and 2019 were included. The rs53576 was the only OXTR gene polymorphism analyzed in all studies. The OXTR genetic variation explained significant changes in N1, P2, N2, P3, and late positive potential (LPP) components during social perception and empathy for pain tasks. OXTR genotypes were not related to P1, N170, N3, or any neural activity after 600 ms. The discussion is focused on the influence of OXTR genetics on neural processing, the development of brain neural networks implicated in social and emotional skills, cultural neuroscience of the oxytocinergic system, and methodological issues of this field. In conclusion, the evidence supports the hypothesis that genetic variations of the OXTR significantly influence neural activity related to emotional and social processing, except for the early phases of face recognition.

Keywords: oxytocin, OXTR gene polymorphism, event-related potentials, social skills, affective processing

1. Introduction

Oxytocin is a neuropeptide that is implicated in relational phenomena such as maternal behaviors, social bonding, emotional communication, caring for others, and social cognition [1, 2]. Further, thanks to evidence from research involving nonhuman animal models [3], intranasal administration [3, 4], developmental psychopathology [5], genetic variations [6], and translational approaches [7], we now know that oxytocin supports the development of socioemotional skills such as perception of social cues, biobehavioral synchrony, regulation of emotions, prosocial concern, perspective taking, and empathy.

In the last decades, researchers have grown aware of the importance of oxytocin receptor (OXTR) gene polymorphisms in explaining individual and demographic variations in the development of social behaviors [6]. All OXTR polymorphisms identified thus far are single nucleotide polymorphisms (SNP), with the most studied SNP being a guanine (G) to adenine (A) substitution in the locus rs53576. These SNPs have been repeatedly associated with differences in sensitivity to social cues, regulation of stress response, social cognition, and brain activity during social tasks [8, 9]. However, some recent research has failed to find differences in emotional traits between carriers of the alleles for this polymorphism. However, some recent research has not found differences in emotional traits between carriers of the alleles of this polymorphism [10], which raises the question of whether genetic variations in the OXTR gene lead to variations in the brain's processing of socioemotional signals.

Event-related potentials (ERPs) may lend a hand to this endeavor. ERPs are used as a reliable and safe method to examine the human neurophysiological activity related with psychological processing [11]. ERPs are changes in the electroencephalogram of a person, which are time-locked to cognitive, motor, affective, and social events presented during rigorously controlled tasks. ERPs require meticulous experimental control, making it easier to establish a fine-grained analysis of the time dynamics of the brain during the processing of specific aspects of stimuli, which can limit the establishment of broader functional analyses. However, it is justified to inquire about the ERP as it is one of the methodological approaches with the most outstanding reputation within cognitive neurosciences since it allows consolidating in greater detail how our brain constructs psychological phenomena [12].

ERPs are analyzed in terms of components that can be studied as a window into the neurophysiological mechanisms associated with ongoing mental activity. These components are built by using the polarity, latency, and scalp distribution of ERP waveforms and by considering the specific conditions of each experimental task [13, 14]. For example, in socio-affective tasks, frequently analyzed components include N170, early posterior negativity (EPN), and late positive potential (LPP). The first is an inferior negative wave, appearing approximately 170 ms after a face is shown; the second is occipital-parietal activity occurring around 200 ms after the display of images with affective content; and the last is anterior brain activity appearing shortly after 500 ms following the perception of arousing stimuli.

Accordingly, it is expected that carriers of distinct OXTR polymorphisms exhibit differences in neural processing, as measured by ERPs, during diverse psychological tasks. However, a systematic review of the studies that have examined the relationship between OXTR polymorphisms and ERP components has not been conducted so far. Such a review is valuable because it will allow for a better understanding of how genetic variations in the oxytocinergic system can lead to neurophysiological and behavioral differences. Additionally, it will be useful in guiding future investigations by identifying points of interest and questions to be answered. Therefore, our main aim is to review systematically the literature that has studied the relationship between OXTR polymorphisms and ERP components elicited during psychological tasks.

2. Method

This review was performed in accordance with the 2020 PRISMA guidelines [15]. **Figure 1** shows a summary of all the procedures that led to the final selection of reviewed reports. Initially, 473 documents were found. Of them, eight studies fulfilled all selection criteria and were reviewed. Described below is each stage of the review process.

2.1 Eligibility criteria

Eligible studies had to be scientific reports that inquired into the relationships between polymorphisms of the OXTR gene and ERPs components in human participants. All reports are needed to describe genotyping procedures, EEG acquisition procedures, ERP analyses, and statistical procedures. Considered studies had to be

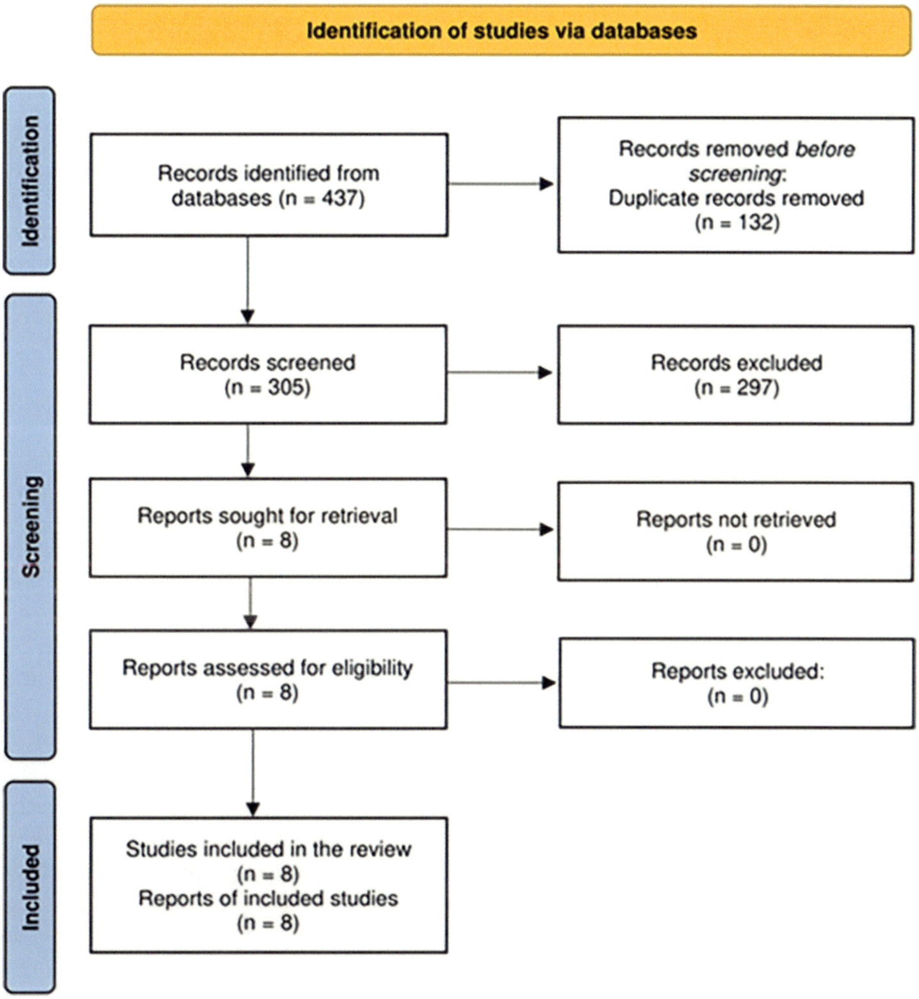

Figure 1.
This PRISMA flow chart depicts the stages of the review process.

published or be in press anytime until April of 2023 in the English or the Spanish languages. No setting or time frame restrictions were imposed. Case series, case reports, conceptual literature, and literature not accomplishing the above criteria were excluded.

2.2 Search strategy

Identification, screening, and inclusion of studies were performed following the 2020 PRISMA guidelines (**Figure 1**). We used the Web of Science search engine to search the Web of Science Core Collection, Medline, the BIOSIS Citation Index, the Korean Journal, the Russian Science Citation Index, and the SciELO Citation Index databases. The search keywords were obtained from previous reports in the OXTR and ERP fields and from the MeSH thesaurus. These keywords were grouped into two lists. The OXTR list was composed of the terms [OXTR], [oxytocin receptor], [oxytocin polymorphism], [oxytocin single-nucleotide polymorphism], [oxytocin single nucleotide polymorphism], and [oxytocin SNP]. The ERP list was composed of the terms [event-related potential], [event related potential], [ERP], [electroencephalography], [EEG], [evoked potential].

Database searches were performed using the following query strategy: ([“OXTR list term 1”] OR ... [“OXTR list term n”]] AND [[“ERP list term 1”] OR ... [“OXTR list term n”]). A less restrictive query string (without enclosing the terms between quotation marks) was also used to confirm no relevant document was left out: ([OXTR list term 1] OR ... [OXTR list term n]] AND [[ERP list term 1] OR ... [OXTR list term n]). The terms had to be mentioned in the titles, abstracts, or keywords. Finally, we set the parameters of the searches to exclude books, conference reports, case reports, and editorials.

2.3 Studies selection and data collection

We read the titles and abstracts of all identified articles for evidence that they could meet the eligibility criteria. The reports that seemed to meet such criteria or could not be regarded as not meeting such criteria were further screened in full to verify inclusion criteria. Once fully scanned, the reports that indeed met the eligibility criteria were included in the review. Next, each article was read thoroughly, and relevant data was copied into a worksheet to be grouped with similar data from other studies. Variables for which data were extracted sought can be organized into five groups: report characteristics, sample characteristics, OXTR genotyping, ERPs and paradigms, and main findings (**Table 1**).

In the report characteristics, data was sought for publication year, country in which the study was carried out, study design, and general topic of research. In the sample characteristics, data was sought for sample size, age, sex, country, and clinical diagnoses. In relation to OXTR genotyping, data was sought for OXTR polymorphisms, genotyping method, tissue for DNA extraction, other genes under study, genotypic groups, allelic frequencies, genotypic frequencies, and Hardy–Weinberg equilibrium (HWE). Regarding ERPs and paradigms, data was sought for preprocessing procedures, characteristics of ERP components (number, family, latency range, number of electrodes, and region of interest), procedures used for ERP calculation, and experimental paradigms. Finally, the main findings were extracted and summarized. We focused solely on the results addressing ERP component comparisons between OXTR genotypic groups, so other results were not analyzed.

DOI: http://dx.doi.org/10.5772/intechopen.112631

Paper	Country	Sample size	Participants	OXTR SNPs	Genotypic groups	EEG sensors	Components	ERP paradigm	Main result
Sjaarda et al. [16]	Canada	167	ASD children and their parents	rs2254298, rs53576, rs7632287, rs1042778	GG, GA, AA	128	P1, N170, P300, N400	Face processing	Nonsignificant results.
Slane et al. [17]	USA	48	School children	rs53576, rs237897, rs1042778, rs2254298	GG, GA, AA	32	N170	Face processing	Nonsignificant results.
Choi et al. [18]	Japan	88	Healthy adult students	rs53576	GG, GA, AA	64	N1, N2, LPP	Affective image processing	Higher N1 and N2 amplitudes to human affective images in GG carriers than AA carriers.
Munk et al. [19]	Germany	150	Young adults	rs53576	GG vs. A+	32	N170	Face processing	Shorter N170 latency to upright angry faces in the right hemisphere of A carriers, but not in GG carriers.
Luo et al. [20]	China	48	Healthy adult students	rs53576	GG vs. AA	64	N1, P2, N2, P3	Empathy for pain	Higher P3 amplitude to sadistic painful faces in GG carriers than AA carriers.
Luo et al. [21]	China	50	Healthy adult students	rs53576	GG vs. AA	62	N1, P2, N2, P3	Empathy for pain	Higher P2 amplitude to suffering ingroup faces in, but not in AA carriers.
Peltola et al. [22]	Finnish	94	Healthy mothers and nonmother adult students	rs53576	GG vs. A+	21	N1, N170, EPN, LPP	Face processing	Shorter N1 latency to strong-intensity infant faces in GG carriers, but not in A carriers.
Fowler et al. [23]	USA	37	Young adults	rs53576	GG vs. A+	40	LPP	Affective image processing	Higher LPP amplitude to aversive images in GG carriers than A carriers.

Table 1.
Summary of relevant findings.

3. Results

3.1 Report characteristics

Eight studies were included in the final review; a summary of these papers is shown in **Table 1**. They were published between 2014 and 2019 in the English language. Two studies were conducted in the United States, two in China, one in Canada, one in Japan, one in Germany, and one in Finland. There was no study from South America, Central America, Africa, Eastern Europe, Asia (other than East Asia), or Oceania.

In general, researchers examined whether OXTR polymorphisms were related to neural differences during socioemotional processing (**Table 1**). Four studies examined the differences between OXTR genotypic groups in ERPs during face processing tasks; two studies looked for associations between ERPs elicited by perception of others' pain tasks and OXTR genotypes; and the two remaining studies were concerned with the differences in neural activity between genotypic groups during emotional cue processing.

3.2 Sample characteristics

Sample size was 85 people on average, with only two investigations engaging over a hundred participants. Only one study mentioned its sample source, which was individuals with autism and their relatives [16]. Two studies recruited participants from both sexes [16, 17], two recruited only males [18, 19], three selected only females [20–22], and one did not state the sexes of the participants [23]. Most of the participants were young adults [75%], while two studies included school-aged children (7–12 years) [16, 17]. No sample included infants, toddlers, preschoolers, adolescents, or elderly (**Table 1**).

3.3 OXTR genotyping

Besides the OXTR gene, one study examined genes linked with monoaminergic neurotransmission, such as DR1, COMT, and SLC6A4 [16]; while another explored the CD38 [19], a gene implicated in oxytocin release. For the OXTR gene, polymorphisms rs53576, rs237897, rs1042778, and rs2254298 were studied, being rs53576 analyzed in all studies. Only one study explored the interactions between different OXTR SNPs and although they did not find any significant effect on ERP components, they did on social cognition measures [17].

The genotypic frequencies for the rs53576 variants across the reviewed studies are shown in **Figure 2**. Some studies did not report the genotypic or allelic frequencies [19, 20, 23], so we calculated these values from their HWE report. The three studies conducted with Asian participants revealed lower GG genotype frequencies, while two of the five studies with Caucasian participants found a greater ratio of GG homozygotes. The three remaining studies did not satisfy the HWE probe, pointing to a large proportion of heterozygotes (**Figure 2**).

3.4 ERPs and paradigms

We detected large discrepancies between the ERP procedures used across the studies. EEG acquisition and ERP preprocessing protocols varied widely in terms of

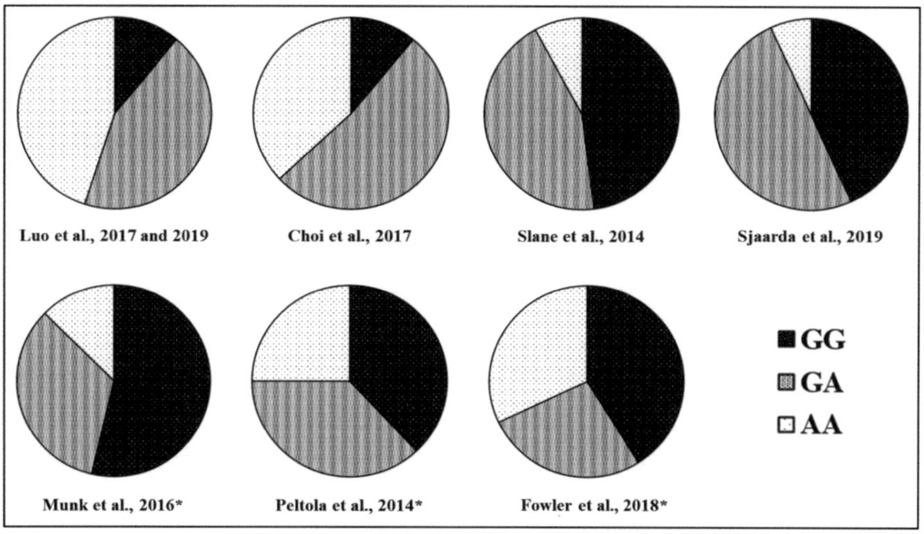

Figure 2.
*Report of genotypic frequencies for rs53576 polymorphism in each study. GG = carriers of the GG genotype; GA = carriers of the GG genotype; AA = carriers of the AA genotype. *Studies that did not satisfy the HWE.*

sensor number (**Table 1**), referencing, filtering, artifact detection, artifact correction, and epochs selection. For example, the four studies that ran algorithms to correct artifacts did not justify their decision. EPR construction was also heterogeneous across the reports. Indeed, the reports generally provided little justification about the procedures that led to electrode number reduction, selection of latency ranges, and potential measurements. Most authors remarked that they selected latency ranges by means of visual inspection of the grand average waveform, and even one study specified a range for each participant [16]. Therefore, latencies for each family of components varied greatly (see component lines in **Figure 3**). Regarding the ERP measures, four studies calculated the latency peak and the amplitude peak for each latency period [16, 17, 19, 22], while the other four studies used the amplitude average for each latency window [18, 20, 21, 23]. All paradigms were time-locked with the stimuli, meaning no paradigm was time-locked with the responses. All studies employed visual affective or visual social stimuli; four used faces, two used affective images, and two used pictures of people suffering (**Table 1**).

3.5 Main findings

Figure 3 summarizes the ERP component amplitude differences between OXTR rs53576 genotypic groups, as reported by the reviewed studies. On the one hand, OXTR genotypes were not significantly associated with the P1 or the N170 components during the socio-affective tasks. Similarly, neuronal responses occurring after 600 ms were not explained by any genetic variation of the OXTR. On the other hand, the studies together found the N1, P2, N2, P3 and LPP potentials to have a significantly higher amplitude in GG carriers during the experimental tasks. Such genotype-dependent components can be grouped into three general time ranges of the neural dynamics (colored heads in **Figure 3**). The earliest differences in neural activity were found between 50 and 200 ms in centro-parietal and frontal zones (respectively N1 and P2; blue head). The OXTR also affected middle neural processing in the latency

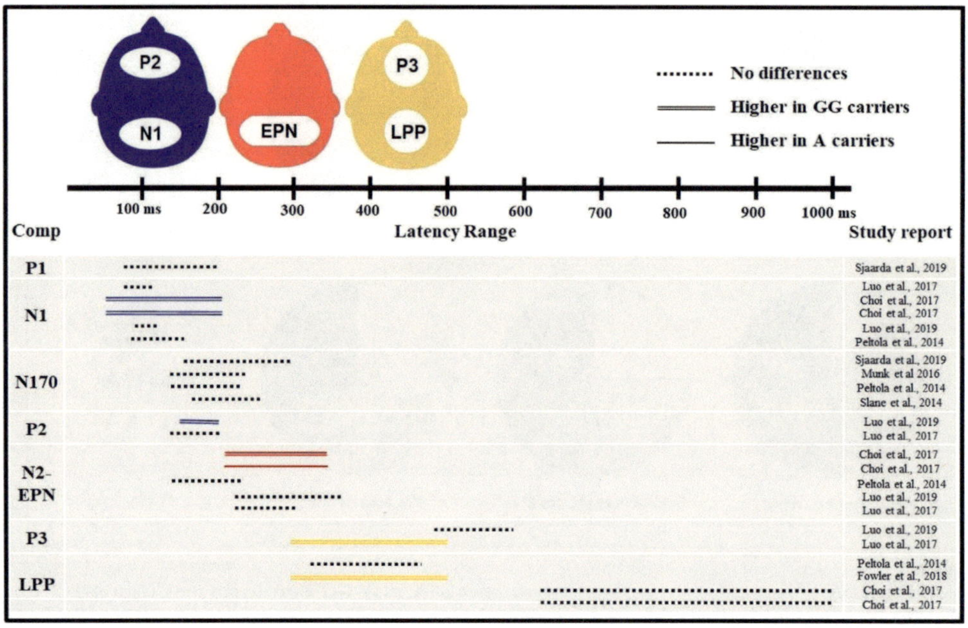

Figure 3.
Differences in ERP components across carriers of distinct OXTR rs53576 polymorphisms. At the top, the heads display the brain regions where components were recorded. Their colors represent a family component: blue = early, red = middle, and orange = late. At the bottom, the leftmost column contains the studied components. The middle column shows latencies in milliseconds, where lines represent the components examined in their respective paper and their lengths indicate the latency range used to calculate the component. Black dotted lines represent nonsignificant differences between genotypes, whereas colored solid lines point out significant differences. Doble lines indicate a higher component amplitude in GG carriers, while single lines denote a higher component amplitude in A-allele carriers. The rightmost column shows the source study.

range between 200 to 350 ms in posterior regions, specifically, the EPN component (red head). The last differences in neural activity were observed in the central regions of both frontal and parietal lobes in latencies between 300 and 500 ms (P3 and LPP; orange head).

Face processing was generally not different across OXTR genotypes. Specifically, amplitudes of neural responses to upright or inverted faces did not vary significantly in healthy adults [19], mothers [22], children [17], or autistic children [16]. However, a significant interaction was observed in the N1 latency. Peltola's team found that, unlike A carriers, GG carriers had a shorter N1 latency to strong-intensity infant faces than they had to mild-intensity ones [22]. Likewise, Munk et al. [19] found a significant interaction in the N170 latency, where A-allele carriers had shorter latencies in the right hemisphere in reaction to upright angry faces. Nevertheless, such a difference was not observed in a replication sample [19].

The processing of affective images varied in function of the rs53576 genotypes. GG homozygotes exhibited a more negative posterior N1 (50–200 ms) during the perception of affective images of humans and objects and a more negative posterior N2 (200–320 ms) when they perceived affective images of humans only [18]. In the same study, no effects of OXTR genotypes on LPP (600–1000 ms) were found. However, Fowler and colleagues found independently that GG participants responded with a higher parietal LPP amplitude (at 300–500 ms) during the perception of aversive pictures. That said, it should be noted that the LPP latencies and regions of interest were considerably different between the two studies, making it difficult to compare the results (**Figure 3**).

Neural responses to the suffering of others were significantly mediated by OXTR genotypes. Specifically, GG carriers exhibited a heightened frontal P2 amplitude (at 136–176 ms) when they were shown painful expressions adopted by people that the participants perceived as themselves [21]. Additionally, in a previous study [20], frontal P3 was found to be more positive for sadistic painful stimuli in the GG group, which may indicate a bigger excitatory activity in the amygdala and insula.

4. Discussion

4.1 OXTR polymorphisms and neural processing

Our results provide evidence that variations in the OXTR gene are associated with differences in brain activity during socioemotional tasks (**Table 1**). These differences were distributed along several time intervals and brain regions (**Figure 3**). In particular, the main differences were found in rs53576 GG homozygotes, who produce wider neural potentials in the face of salient social stimuli such as people expressing pain.

GG homozygotes exhibited more negative early activity (N1) in posterior cerebral regions when they perceived socially salient cues (**Figure 3**). Psychophysiological studies have previously linked enhancement of the posterior N1 with an attentional shift to prominent stimuli and implicit visual discrimination [24, 25]. Therefore, GG carriers may detect and recognize socio-affective signals more readily. This enhanced activity may originate in interneurons in the occipital, parietal, and temporal areas implicated in social perception. Additionally, it is possible that these areas are excited by afferences from projection neurons in the limbic area, which is rich in OXTRs.

GG homozygotes also had stronger early anterior positive potentials (P2) when they watched people suffering (**Figure 3**). This early frontal excitation may be associated with the involuntary allocation of attentional resources to the processing of the highly arousing images [26, 27], suggesting GG carriers may bear neural mechanisms that enable them to have a more sensitive perception of emotional signals from others.

Moreover, OXTR SNPs produced variations in the N2-EPN potentials in posterior brain regions, with GG homozygotes having wider negative potentials to affective human pictures and AG-AA carriers doing so to affective pictures without humans (**Figure 3**). The N2-EPN component is a typical index of attentional engagement to emotional pictures [28, 29]. It may be generated from interneuron activity in brain regions such as the amygdala, the hippocampus, and the superior and inferior parietal lobes [30, 31], which contain an abundance of OXTRs [32, 33]. Therefore, it seems that OXTR rs53766 polymorphisms influence the activity of posterior corticolimbic networks, which enhances GG carriers' and AA-AG carriers' discrimination and categorization of social cues and nonsocial arousing stimuli, respectively.

The late components modulated by the OXTR rs53576 polymorphisms were positive potentials at frontal (P3; **Figure 3**) and parietal sites (LPP; **Figure 3**). Both ERP components were higher in GG carriers when they perceived pain faces and aversive pictures. The P3 is usually interpreted as an indicator of effortful decision-making during demanding tasks [34, 35], as was the case in Luo's experiment [20]. The LPP component is elicited by highly arousing stimuli and has been associated with enhanced memory encoding, appraisal, and suppression of response to affective pictures [36, 37]. These findings may indicate that GG homozygotes' greater sensitivity to social images allows them to execute enhanced top-down processes to evaluate arousing social contexts and to control their responses in accordance with such contexts. However, these interpretations are necessarily preliminary, considering that six of the reviewed studies failed to find any significant differences in the LPP component between OXTR genotypes.

Overall, these findings are consistent with the social salience hypothesis, which contends that an increase in oxytocinergic neurotransmission influence neural activity in such a way that the processing of social information is enhanced [38, 39]. Animal research has proven that SNPs in the noncoding region of the OXTR gene have an impact on the number of OXTR receptors in brain regions implicated in social behaviors [40]. Likewise, neuroimage studies have confirmed that OXTR SNPs are associated with variations in brain activity in areas responsible for the processing of social information [8, 41]. In brief, GG homozygotes for the rs53576 SNP may have a higher density of receptors in specific critical areas, which may enhance oxytocinergic neurotransmission in neural networks implicated in the processing of social cues [40].

The null results on the face-specific N170 component are surprising because there exists a long-standing theoretical association between oxytocinergic neurotransmission and face processing [42]. Indeed, Skuse et al. found a link between the OXTR rs237887 alleles and face recognition in families with autistic children [43]. Moreover, fMRi studies have also found significant effects on face recognition involving OXTR SNPs. For example, the Westberg team reported that rs7632287 genotypes differ in recognition of faces and amygdala activity [44]. Similarly, O'Connell and coworkers linked the rs2268498 SNP with inferior occipital gyrus activity due to perception of fear expressions [45]. Nevertheless, these results should be analyzed cautiously given that the Skuke study used a very particular sample and the fMRi studies used tasks sensitive to other psychological functions such as emotional processing and mental inference. What is more, the lack of association between OXTR SNPs and the N170 component, is in line with the nonsignificant relationship between many OXTR SNPs and several face recognition tasks, which was found in an exhaustive study by [46]. Therefore, this evidence together seems to indicate that oxytocinergic neurotransmission is not essential to the early stages of discrimination of facial configurations.

Furthermore, these results could help to understand the conflicting findings on the relationship between genetic variations in the oxytocinergic system and human psychological phenotypes [10]. Variations between genotypes in neural processing during exposure to social cues with high emotional content observed during intermediate and late latencies (100–600 ms) indicate that the oxytocinergic system is central in facilitating the implicit processing of emotional signs during social interactions. Still, this system would have a less relevant and direct role in the awareness of one's affective states; therefore, a minimal effect could be expected when self-report questionnaires or verbal responses are used. In this sense, future studies would benefit from using experimental tasks instead of questionnaires and psychological tests to assess social and emotional phenotypes.

4.2 OXTR polymorphism, neural functioning, and human development

OXTR polymorphisms influence brain development by modulating the density of OXTRs, by modifying the sensibility toward the social environment, and by orchestrating fine-tuned social transactions during critical periods of brain maturation. Consequently, OXTR rs53576 GG individuals tend to be more sensitive to their social environment, developing larger phenotypic variability [47]. For instance, in

G allele carriers, protective and synchronic caring favors the development of better emphatic, prosocial, and emotional skills, while childhood adversity leads to more avoidant behaviors and poor social skills. In contrast, A carriers, who are less sensitive, show fewer developmental variations [48, 49]. Moreover, effects of OXTR SNPs on the modulation of developmental plasticity have also been found in neuroimage studies, where limbic and frontal networks have been shown to be more plastic [8, 50, 51]. Plasticity in fronto-limbic networks is consistent with the differences in the ERP components observed in this review.

The developmental plasticity associated with some OXTR genotypes may be due to epigenetic mechanisms. Indeed, numerous studies have reported an association between OXTR DNA methylation and differences in social cognition, emotional behaviors, and neuroendocrine functioning in several moments of human and animal development [52–55]. A recent systematic review found that an increase in OXTR gene methylation was linked with a reduction in receptor expression, social sensitivity, and developmental plasticity, leading to poor social skills and affective dysregulation in healthy and psychopathologic samples [56]. A leading hypothesized mechanism is that people with more G alleles have more CpG islands, facilitating epigenetic modulation and major developmental plasticity, with early adverse experiences as the main predictor of OXTR hypermethylation.

4.3 OXTR polymorphism, neural functioning, and cultural differences

Figure 2 shows a large variation in the allelic frequencies of the OXTR rs53576 SNP between Caucasian and Asian samples, which is in line with previous findings that Asian populations have higher A-allele frequencies [57]. This geographic distribution of OXTR genotypes is associated with cultural patterns, including collectivistic values, control of emotional expressions, emotional support seeking, social interdependence, empathy for pain, altruism motivation, prevalence of depression, and brain functioning [58–60]. These cultural and genetic differences may have been shaped in human societies throughout history. Selection for A alleles in collectivistic nations could be the result of son favoritism, prolonged infanticide, and marriage patterns, whereas G allele accumulation may have been facilitated by mothers' investment in childcare and male cooperation in everyday life activities.

These cultural and demographic factors should be considered to better interpret neural functioning differences. In the first place, genotypic frequencies may limit statistical analysis. For example, since Caucasian samples have fewer AA carriers, such studies require larger samples to find significant results. Further, as cultural values and relational tendencies have important effects on various genotypes, it is indispensable to include measures for these variables. Finally, samples that do not satisfy the HWE must be carefully analyzed because the sampling could be biased, or evolutionary pressures could be affecting the genetics of these populations [61].

4.4 Limitations and future directions

4.4.1 Samples

Three important sample issues can be identified. First, no studies included participants from South America, Central America, Africa, Eastern Europe, the Middle East, or Oceania. There exist important variations in the allelic frequencies between these regions, which may be related to differences in behavior and brain function. Therefore, generalization of these results to diverse geographic and cultural regions is limited, such as in Conner et al. [10] report, who did not find an association between OXTR rs53576 SNP and emotional trait, but they only included a sample from New Zealand. Second, studies have focused on young adults, and there are no inquiries on infants, adolescents, or elderly. Since adolescence is considered a sensitive period for developing social skills [62], the lack of studies at this age range restricts our understanding of how OXTR polymorphisms influence the development of brain activity and social behavior from childhood to adulthood. There is evidence that during adolescence OXTR polymorphisms produce high social sensitivity, opening a critical period to rewiring brain networks and reorganizing behavior, as the oxytocin system interacts with pubertal hormones to create age- and sex-dependent developmental trajectories [63]. And third, the small sample sizes used in the studies limit the possibility of finding significant results. Guidelines to ERPs recommend including over 40 participants per condition [64], meaning researchers may need to incorporate samples of more than 120 participants to analyze the effect of variables such as sex, age, and other polymorphisms. Finally, larger samples are necessary to get lower p-values and more marked effect sizes, which are ideal conditions to reproduce results and consolidate this scientific field, although if large samples of different populations and cultural contexts are included, so much genetic and phenotypic variation can be added that it can be challenging to get satisfactory results.

4.4.2 Sex differences

None of the studies showed sex-dependent effects of OXTR polymorphisms on the ERP components, which was unexpected considering previous findings involving OXTR polymorphism–sex interaction effects on behavior and brain function [65]. For the OXTR rs53576 SNP, there is evidence of sex-dependent effects on the volume of limbic structures, such as the hypothalamus and the amygdala [41], and on functional connectivity of the prefrontal cortex [66, 67]. Furthermore, it has been asserted that sex-dependent differences in the development of brain function and social behavior across different OXTR genotypes are likely linked to the role of sexual hormones regulating OXTR expression [68, 69]. It is likely that, as sample sizes increase, sex-dependent effects will be detected.

4.4.3 ERP procedures

As all studies were interested in behavioral tasks tapping social or emotional processing (**Table 1**), there were no inquiries about ERP components for other psychological functions such as object perception, attention, memory, language, executive functioning, and motor control. This is a major gap to fill, as behavioral and fMRI studies have shown that OXTR polymorphisms may partly explain the differences in the development of these cognitive processes [6, 70].

The experimental paradigms used in the studies hinder comparison between ERPs components such as P3, Nc, LPP, slow wave potentials, and error-related negativity among OXTR genotypes. In relation to EEG acquisition and preprocessing, we detected wide variation in the number of electrodes, referencing, filtering, sampling rate, artifact detection, elimination, and correction, which makes it hard to compare ERPs results [71]. In general, the reports lacked substantial explanations about the selection of ERP measurements, latency ranges, electrode reduction methods, and statistical procedures. Most researchers used visual inspection of the grand average to select latency ranges, employed amplitude, and latency peaks and chose electrodes autonomously to calculate each component, being all these procedures discouraged by ERP guidelines [27, 72].

4.4.4 Publication bias

Our objective was to carry out a systematic review but not a meta-analysis because of the heterogeneity in ERP procedures, psychological paradigms, statistics for hypothesis testing, and number of subgroups compared. All these make very difficult to run quantitative analyses for publication bias, sensitivity, and subgroups. However, in the exploratory analysis we did not find publication bias solid evidence; three of the eight studies report negative results with nonsignificant differences between OXTR genotypes. Moreover, of the 23 components analyzed (**Figure 3**), seven [30%] showed significant differences between genotypes, and in the 70% of comparison, the differences were nonsignificant. However, small sample sizes (**Table 1**), small effect sizes, and high p-value (<0.05) in two studies with positive results could mean a possible bias in these publications [22, 23]. In the future, editors could demand higher effect sizes and narrower confidence intervals in the reports of quantitative results, as suggested in the APA guidelines, which may facilitate the comparison of studies.

In short, future studies will benefit from the use of larger samples, more heterogenic aged populations in the samples, and the recruitment of participants from different geographical places around the world. Additionally, it will be helpful for studies to include sex or gender as a control variable, especially when the samples include adolescents. Also, it may be advantageous for the field to demand that studies use reliable, conventional, and standardized protocols to acquire and process EEG signals. Accumulating evidence from future studies will help to establish, refute, and clarify how EPR components function as a window into the oxytocinergic processing of human psychological functions.

5. Conclusion

In this systematic review, we have found suggestive, preliminary evidence supporting the hypothesis that the rs53576 polymorphism of the OXTR gene significantly influences ERP components elicited during socioemotional tasks. Prominently, larger N1, P2, EPN, P3, and LPP components in GG homozygotes seem to be associated with an increase in the sensitivity to salient social cues in social and emotional situations. Moreover, genetic variations in the OXTR do not affect the neural activity during the earlier moments of perceiving faces.

Acknowledgements

We thank the *Universidad de los Andes* and the *Universidad Externado de Colombia* for their support in the publication of this work.

Conflict of interest

The authors declare no conflict of interest.

Author details

Diego Armando León-Rodríguez[1,2*], Julian Becerra[2], Juan Carlos Caicedo Mera[1], Luis Fernando Cardenas[2*], Jorge Martínez Cotrina[1] and Diego Mauricio Aponte Canencio[1]

1 Facultad de Ciencias Sociales y Humanas, Universidad Externado de Colombia, Bogotá, Colombia

2 Departamento de Psicología, Universidad de los Andes, Bogotá, Colombia

*Address all correspondence to: diego.leon4@uexternado.edu.co
and lucarden@uniandes.edu.co

References

[1] Bosch OJ, Young LJ. Oxytocin and social relationships: from attachment to bond disruption. Current Topics in Behavioral Neurosciences. 2018;**35**:97-117. DOI: 10.1007/7854_2017_10

[2] Ebert A, Brüne M. Oxytocin and social cognition. Current Topics in Behavioral Neurosciences. 2018;**35**: 375-388. DOI: 10.1007/7854_2017_21

[3] Althammer F, Jirikowski G, Grinevich V. The oxytocin system of mice and men - similarities and discrepancies of oxytocinergic modulation in rodents and primates. Peptides. 2018;**109**:1-8. DOI: 10.1016/j.peptides.2018.09.003

[4] Quintana DS, Lischke A, Grace S, Scheele D, Ma Y, Becker B. Advances in the field of intranasal oxytocin research: lessons learned and future directions for clinical research. Molecular Psychiatry. 2021;**26**(1):80-91. DOI: 10.1038/s41380-020-00864-7

[5] Guastella AJ, Hickie IB. Oxytocin treatment, circuitry, and autism: a critical review of the literature placing oxytocin into the autism context. Biological Psychiatry. 2016;**79**(3):234-242. DOI: 10.1016/j.biopsych.2015.06.028

[6] Feldman R, Monakhov M, Pratt M, Ebstein RP. Oxytocin pathway genes: evolutionary ancient system impacting on human affiliation, sociality, and psychopathology. Biological Psychiatry. 2016;**79**(3):174-184. DOI: 10.1016/j.biopsych.2015.08.008

[7] Shamay-Tsoory S, Young LJ. Understanding the oxytocin system and its relevance to psychiatry. Biological Psychiatry. 2016;**79**(3):150-152. DOI: 10.1016/j.biopsych.2015.10.014

[8] Schneider-Hassloff H, Straube B, Jansen A, Nuscheler B, Wemken G, Witt SH, et al. Oxytocin receptor polymorphism and childhood social experiences shape adult personality, brain structure and neural correlates of mentalizing. NeuroImage. 2016;**134**:671-684. DOI: 10.1016/j.neuroimage.2016.04.009

[9] Li J, Zhao Y, Li R, Broster LS, Zhou C, Yang S. Association of oxytocin receptor gene [OXTR] rs53576 polymorphism with sociality: a meta-analysis. PLoS One. 2015;**10**(6):e0131820. DOI: 10.1371/journal.pone.0131820

[10] Conner TS, McFarlane KG, Choukri M, Riordan BC, Flett JAM, Phipps-Green AJ, et al. The oxytocin receptor gene [OXTR] variant rs53576 is not related to emotional traits or states in young adults. Frontiers in Psychology. 2018;**9**:2548. DOI: 10.3389/fpsyg.2018.02548

[11] Luck SJ, Gaspelin N. How to get statistically significant effects in any ERP experiment [and why you shouldn't]. Psychophysiology. 2017;**54**(1):146-157. DOI: 10.1111/psyp.12639

[12] MacNamara A, Joyner K, Klawohn J. Event-related potential studies of emotion regulation: a review of recent progress and future directions. International Journal of Psychophysiology. 2022;**176**:73-88. DOI: 10.1016/j.ijpsycho.2022.03.008

[13] Donoghue T, Voytek B. Automated meta-analysis of the event-related potential [ERP] literature. Scientific Reports. 2022;**12**:1867. DOI: 10.1038/s41598-022-05939-9

[14] Sur S, Sinha VK. Event-related potential: an overview. Industrial

Psychiatry Journal. 2009;**18**(1):70-73. DOI: 10.4103/0972-6748.57865

[15] Page MJ, McKenzie JE, Bossuyt PM, et al. The PRISMA 2020 statement: an updated guideline for reporting systematic reviews. BMJ. 2021;**372**:n71. DOI: 10.1136/bmj.n71

[16] Sjaarda CP, Sabbagh M, Wood S, Ward-King J, McNaughton AJM, Hudson ML, et al. Homozygosity for the 10-repeat dopamine transporter (DAT1) allele is associated with reduced EEG response in males with ASD. Research in Autism Spectrum Disorder. 2019;**60**:25-35. DOI: 10.1016/j.rasd.2018.12.003

[17] Slane MM, Lusk LG, Boomer KB, Hare AE, King MK, Evans DW. Social cognition, face processing, and oxytocin receptor single nucleotide polymorphisms in typically developing children. Developmental Cognitive Neuroscience. 2014;**9**:160-171. DOI: 10.1016/j.dcn.2014.04.001

[18] Choi D, Minote N, Watanuki S. Associations between the oxytocin receptor gene (OXTR) rs53576 polymorphism and emotional processing of social and nonsocial cues: an event-related potential (ERP) study. Journal of Physiological Anthropology. 2017;**36**(1):12. DOI: 10.1186/s40101-016-0125-3

[19] Munk AJ, Hermann A, El Shazly J, Grant P, Hennig J. The idea is good, but...: failure to replicate associations of oxytocinergic polymorphisms with face-inversion in the N170. PLoS One. 2016;**11**(3):e0151991. DOI: 10.1371/journal.pone.0151991

[20] Luo S, Zhu Y, Xu Y, Kong Q. The oxytocinergic system modulates sadistic context-dependent empathy in humans. Scientific Reports. 2017;**7**(1):12463. DOI: 10.1038/s41598-017-12671-2

[21] Luo S, Zhang T, Li W, Yu M, Hein G, Han S. Interactions between oxytocin receptor gene and intergroup relationship on empathic neural responses to others' pain. Social Cognitive and Affective Neuroscience. 2019;**14**(5):505-517. DOI: 10.1093/scan/nsz029

[22] Peltola MJ, Yrttiaho S, Puura K, et al. Motherhood and oxytocin receptor genetic variation are associated with selective changes in electrocortical responses to infant facial expressions. Emotion. 2014;**14**(3):469-477. DOI: 10.1037/a0035959

[23] Fowler JJ, Braley KL, Farero RD, et al. The late positive potential and subjective arousal ratings evoked by negative images vary as a function of oxytocin receptor genotype SNP rs53576. Neuroreport. 2018;**29**(14):1145-1150. DOI: 10.1097/WNR.0000000000001085

[24] Bar-Haim Y, Lamy D, Glickman S. Attentional bias in anxiety: a behavioral and ERP study. Brain and Cognition. 2005;**59**:11-22. DOI: 10.1016/j.bandc.2005.03.005

[25] Bruyns-Haylett M, Luo J, Kennerley AJ, Harris S, Boorman L, Milne E, et al. The neurogenesis of P1 and N1: a concurrent EEG/LFP study. NeuroImage. 2017;**146**:575-588. DOI: 10.1016/j.neuroimage.2016.09.034

[26] Feng C, Li W, Tian T, Luo Y, Gu R, Zhou C, et al. Arousal modulates valence effects on both early and late stages of affective picture processing in a passive viewing task. Social Neuroscience. 2014;**9**(4):364-377. DOI: 10.1080/17470919.2014.896827

[27] Kappenman ES, Luck SJ. Best practices for event-related potential research in clinical populations. Biological Psychiatry: Cognitive Neuroscience and Neuroimaging.

2016;**1**(2):110-115. DOI: 10.1016/j.bpsc.2015.11.007

[28] Schupp HT, Flaisch T, Stockburger J, Junghöfer M. Emotion and attention: event-related brain potential studies. Progress in Brain Research. 2006;**156**:31-51. DOI: 10.1016/S0079-6123[06]56002-9

[29] Schupp HT, Schmälzle R, Flaisch T, Weike AI, Hamm AO. Affective picture processing as a function of preceding picture valence: an ERP analysis. Biological Psychology. 2012;**91**(1):81-87. DOI: 10.1016/j.biopsycho.2012.04.006

[30] Grass A, Bayer M, Schacht A. Electrophysiological correlates of emotional content and volume level in spoken word processing. Frontiers in Human Neuroscience. 2016;**10**:326. DOI: 10.3389/fnhum.2016.00326

[31] Rozenkrants B, Olofsson JK, Polich J. Affective visual event-related potentials: arousal, valence, and repetition effects for normal and distorted pictures. International Journal of Psychophysiology. 2008;**67**(2):114-123. DOI: 10.1016/j.ijpsycho.2007.10.010

[32] Grinevich V, Knobloch-Bollmann HS, Eliava M, Busnelli M, Chini B. Assembling the puzzle: pathways of oxytocin signaling in the brain. Biological Psychiatry. 2016;**79**(3):155-164. DOI: 10.1016/j.biopsych.2015.04.013

[33] Quintana DS, Rokicki J, van der Meer D, Alnæs D, Kaufmann T, Córdova-Palomera A, et al. Oxytocin pathway gene networks in the human brain. Nature Communications. 2019;**10**(1):668. DOI: 10.1038/s41467-019-08503-8

[34] Brown SB, van Steenbergen H, Band GP, de Rover M, Nieuwenhuis S. Functional significance of the emotion-related late positive potential. Frontiers in Human Neuroscience. 2012;**6**:33. DOI: 10.3389/fnhum.2012.00033

[35] Nieuwenhuis S, De Geus EJ, Aston-Jones G. The anatomical and functional relationship between the P3 and autonomic components of the orienting response. Psychophysiology. 2011;**48**(2):162-175. DOI: 10.1111/j.1469-8986.2010.01057.x

[36] Macnamara A, Ochsner KN, Hajcak G. Previously reappraised: the lasting effect of description type on picture-elicited electrocortical activity. Social Cognitive and Affective Neuroscience. 2011;**6**(3):348-358. DOI: 10.1093/scan/nsq053

[37] Olofsson JK, Nordin S, Sequeira H, Polich J. Affective picture processing: an integrative review of ERP findings. Biological Psychology. 2008;**77**(3):247-265. DOI: 10.1016/j.biopsycho.2007.11.006

[38] Bartz JA, Zaki J, Bolger N, Ochsner KN. Social effects of oxytocin in humans: context and person matter. Trends in Cognitive Sciences. 2011;**15**(7):301-309. DOI: 10.1016/j.tics.2011.05.002

[39] Shamay-Tsoory SG, Abu-Akel A. The social salience hypothesis of oxytocin. Biological Psychiatry. 2016;**79**(3):194-202. DOI: 10.1016/j.biopsych.2015.07.020

[40] King LB, Walum H, Inoue K, Eyrich NW, Young LJ. Variation in the oxytocin receptor gene predicts brain region-specific expression and social attachment. Biological Psychiatry. 2016;**80**(2):160-169. DOI: 10.1016/j.biopsych.2015.12.008

[41] Tost H, Kolachana B, Hakimi S, Lemaitre H, Verchinski BA, Mattay VS, et al. A common allele in the oxytocin receptor gene (OXTR) impacts prosocial

temperament and human hypothalamic-limbic structure and function. Proceedings of the National Academy of Sciences of the United States of America. 2010;**107**(31):13936-13941. DOI: 10.1073/pnas.1003296107

[42] Lopatina OL, Komleva YK, Gorina YV, Higashida H, Salmina AB. Neurobiological aspects of face recognition: the role of oxytocin. Frontiers in Behavioral Neuroscience. 2018;**12**:195. DOI: 10.3389/fnbeh.2018.00195

[43] Skuse DH, Lori A, Cubells JF, Lee I, Conneely KN, Puura K, et al. Common polymorphism in the oxytocin receptor gene (OXTR) is associated with human social recognition skills. Proceedings of the National Academy of Sciences of the United States of America. 2014;**111**(5):1987-1992. DOI: 10.1073/pnas.1302985111

[44] Westberg L, Henningsson S, Zettergren A, Svärd J, Hovey D, Lin T, et al. Variation in the oxytocin receptor gene is associated with face recognition and its neural correlates. Frontiers in Behavioral Neuroscience. 2016;**10**:178. DOI: 10.3389/fnbeh.2016.00178

[45] O'Connell G, Whalley HC, Mukherjee P, Stanfield AC, Montag C, Hall J, et al. Association of genetic variation in the promoter region of OXTR with differences in social affective neural processing. Journal of Behavioral and Brain Science. 2012;**2**(1):60-66. DOI: 10.4236/jbbs.2012.21007

[46] Verhallen RJ, Bosten JM, Goodbourn PT, Lawrance-Owen AJ, Bargary G, Mollon JD. The oxytocin receptor gene (OXTR) and face recognition. Psychological Science. 2017;**28**(1):47-55. DOI: 10.1177/0956797616672269

[47] Smearman EL, Almli LM, Conneely KN, Brody GH, Sales JM, Bradley B, et al. Oxytocin receptor genetic and epigenetic variations: association with child abuse and adult psychiatric symptoms. Child Development. 2016;**87**(1):122-134. DOI: 10.1111/cdev.12493

[48] Bonassi A, Cataldo I, Gabrieli G, Foo JN, Lepri B, Esposito G. Oxytocin receptor gene polymorphisms and early parental bonding interact in shaping Instagram social behavior. International Journal of Environmental Research and Public Health. 2020;**17**(19):7232. DOI: 10.3390/ijerph17197232

[49] Truzzi A, Poquérusse J, Setoh P, Shinohara K, Bornstein MH, Esposito G. Oxytocin receptor gene polymorphisms (rs53576) and early paternal care sensitize males to distressing female vocalizations. Developmental Psychobiology. 2018;**60**(3):333-339. DOI: 10.1002/dev.21606

[50] Cataldo I, Neoh MJ, Chew WF, Foo JN, Lepri B, Esposito G. Oxytocin receptor gene and parental bonding modulate prefrontal responses to cries: a NIRS Study. Scientific Reports. 2020;**10**(1):8588. DOI: 10.1038/s41598-020-65582-0

[51] Dannlowski U, Kugel H, Grotegerd D, Redlich R, Opel N, Dohm K, et al. Disadvantage of social sensitivity: interaction of oxytocin receptor genotype and child maltreatment on brain structure. Biological Psychiatry. 2016;**80**(5):398-405. DOI: 10.1016/j.biopsych.2015.12.010

[52] Haas BW, Filkowski MM, Cochran RN, Denison L, Ishak A, Nishitani S, et al. Epigenetic modification of OXT and human sociability. Proceedings of the National Academy of Sciences of the United States of America.

2016;**113**(27):E3816-E3823. DOI: 10.1073/pnas.1602809113

[53] Jack A, Connelly JJ, Morris JP. DNA methylation of the oxytocin receptor gene predicts neural response to ambiguous social stimuli. Frontiers in Human Neuroscience. 2012;**6**:280. DOI: 10.3389/fnhum.2012.00280

[54] Krol KM, Puglia MH, Morris JP, Connelly JJ, Grossmann T. Epigenetic modification of the oxytocin receptor gene is associated with emotion processing in the infant brain. Developmental Cognitive Neuroscience. 2019;**37**:100648. DOI: 10.1016/j.dcn.2019.100648. Epub 2019 Apr 26

[55] MacKinnon AL, Feeley N, Gold I, Hayton B, King L, Nagy C, et al. The interaction between oxytocin receptor gene methylation and maternal behavior on children's early theory of mind abilities. Development and Psychopathology. 2020;**32**(2):511-519. DOI: 10.1017/S0954579419000257. PMID: 31030686

[56] Maud C, Ryan J, McIntosh JE, Olsson CA. The role of oxytocin receptor gene (OXTR) DNA methylation (DNAm) in human social and emotional functioning: a systematic narrative review. BMC Psychiatry. 2018;**18**(1):154. DOI: 10.1186/s12888-018-1740-9

[57] Butovskaya PR, Lazebny OE, Sukhodolskaya EM, Vasiliev VA, Dronova DA, Fedenok JN, et al. Polymorphisms of two loci at the oxytocin receptor gene in populations of Africa, Asia and South Europe. BMC Genetics. 2016;**17**:17. DOI: 10.1186/s12863-015-0323-8

[58] Kim HS, Sherman DK, Sasaki JY, Xu J, Chu TQ, Ryu C, et al. Culture, distress, and oxytocin receptor polymorphism (OXTR) interact to influence emotional support seeking. Proceedings of the National Academy of Sciences of the United States of America. 2010;**107**(36):15717-15721. DOI: 10.1073/pnas.1010830107. Epub 2010 Aug 19

[59] Kim MJ, Loucks RA, Palmer AL, Brown AC, Solomon KM, Marchante AN, et al. The structural and functional connectivity of the amygdala: from normal emotion to pathological anxiety. Behavioural Brain Research. 2011;**223**(2):403-410. DOI: 10.1016/j.bbr.2011.04.025. Epub 2011 Apr 22

[60] Luo S, Ma Y, Liu Y, et al. Interaction between oxytocin receptor polymorphism and interdependent culture values on human empathy. Social Cognitive and Affective Neuroscience. 2015;**10**(9):1273-1281. DOI: 10.1093/scan/nsv019

[61] Trikalinos TA, Salanti G, Khoury MJ, Ioannidis JP. Impact of violations and deviations in Hardy-Weinberg equilibrium on postulated gene-disease associations. American Journal of Epidemiology. 2006;**163**(4):300-309. DOI: 10.1093/aje/kwj046

[62] Fuhrmann D, Knoll LJ, Blakemore SJ. Adolescence as a sensitive period of brain development. Trends in Cognitive Sciences. 2015;**19**(10):558-566. DOI: 10.1016/j.tics.2015.07.008

[63] Tereshchenko S, Kasparov E, Zobova S, Smolnikova M, Evert L, Semenova N, et al. Oxytocin pathway gene (CD38, OXTR) variants are not related to psychosocial characteristics defined by strengths and difficulties questionnaire in adolescents: a field school-based study. Frontiers in Psychiatry. 2021;**12**:714093. DOI: 10.3389/fpsyt.2021.714093

[64] Clayson PE, Carbine KA, Baldwin SA, Larson MJ. Methodological

reporting behavior, sample sizes, and statistical power in studies of event-related potentials: barriers to reproducibility and replicability. Psychophysiology. 2019;**56**(11):e13437. DOI: 10.1111/psyp.13437

[65] Stankova T, Eichhammer P, Langguth B, Sand PG. Sexually dimorphic effects of oxytocin receptor gene (OXTR) variants on harm avoidance. Biology of Sex Differences. 2012;**3**(1):17. DOI: 10.1186/2042-6410-3-17

[66] Wang J, Qin W, Liu B, Wang D, Zhang Y, Jiang T, et al. Variant in OXTR gene and functional connectivity of the hypothalamus in normal subjects. NeuroImage. 2013;**81**:199-204. DOI: 10.1016/j.neuroimage.2013.05.029

[67] Wang J, Zhang Y, Zhu D, Yang Z, Liu F, Qin W, et al. A common variant in OXTR rs53576 impacts topological patterns of brain functional networks. European Child & Adolescent Psychiatry. 2020;**29**(7):993-1002. DOI: 10.1007/s00787-019-01414-5

[68] Armeni AK, Assimakopoulos K, Marioli D, et al. Impact of estrogen receptor α gene and oxytocin receptor gene polymorphisms on female sexuality. Endocrine Connections. 2017;**6**(1):44-52. DOI: 10.1530/EC-16-0090

[69] Murata T, Murata E, Liu CX, Narita K, Honda K, Higuchi T. Oxytocin receptor gene expression in rat uterus: regulation by ovarian steroids. The Journal of Endocrinology. 2000;**166**(1):45-52. DOI: 10.1677/joe.0.1660045. PMID: 10856882

[70] Hiraoka D, Nomura M. Maternal childhood adversity, OXTR genotype and cognitive load impact on perceptual and behavioral responses to infant crying. Psychoneuroendocrinology. 2019;**104**:195-202. DOI: 10.1016/j.psyneuen.2019.03.005

[71] Luck SJ. An Introduction to the Event-Related Potential Technique. An Introduction to the Event-Related Potential Technique. Cambridge, MA: MIT Press; 2014

[72] Bigdely-shamlo N, Mullen T, Kothe C, Su K-M, Widmann A, Bigdely-Shamlo N, et al. The PREP pipeline: standardized preprocessing for large-scale EEG analysis. Frontiers in Neuroinformatics. 2015;**9**:16. DOI: 10.3389/fninf.2015.00016. PMID: 26150785

Chapter 6

Oxytocin and Its Congeners in Obstetrics Practice: An Update on Carbetocin

Amit Bhalla and Sandeep Kaushal

Abstract

There are no standardised recommendations for the use of oxytocin in obstetric indications. To prevent postpartum haemorrhage (PPH), the routine administration of oxytocin is standard practice. Failure of prophylactic therapy with oxytocin occurs commonly, necessitating the use of further oxytocin or other treatments to maintain haemodynamic stability. Oxytocin has its limitations as it requires cold storage and transport, and in low-resource settings, the cold chain is not commonly available. By modifying the oxytocin molecule, its half-life has been prolonged and its enzymatic degradation reduced. The modified molecule is named carbetocin. Heat-stable carbetocin is a promising alternative to oxytocin, which can overcome the persistent problems with oxytocin quality as it does not require a cold chain for storage and transport.

Keywords: postpartum haemorrhage, uterotonic, oxytocin, cold chain, heat-stable carbetocin

1. Introduction

Oxytocin is used for induction of labor, augmentation of labor, and to reduce the risk of postpartum haemorrhage (PPH) [1]. Optimal management of oxytocin infusion requires effective interprofessional communication and collaboration. Oxytocin is a peptide generated from the hypothalamus in a pulsatile manner and secreted through the posterior pituitary, stimulating myometrial cells in the uterus and the myoepithelial cells around the mammary alveoli [2]. Oxytocin has a central role in labour. The oxytocin levels gradually increase during pregnancy, and the oxytocin receptors in the uterine muscles also gradually increase in number and become more sensitive to oxytocin in late pregnancy [3, 4]. During the labour initiation, there are fluctuations in the levels of oestrogen and progesterone alongwith changes in their receptors distribution. The effacement of the cervix happens during first stage of labour.

The uterus is supplied by the autonomic nervous system, which has a significant effect on the labour. Parasympathetic stimulation enhances contractility and circulation to the uterus and fetus, while sympathetic activation triggers ineffective contractions and inhibits uterine circulation [5]. Surroundings perceived as safe, familiar and friendly (e.g. a woman's own home) and a supportive environment (e.g. one to one nursing care) are likely to increase oxytocin release by parasympathetic stimulation and facilitate the progression of labour, as well as lead to the beneficial central actions caused by oxytocin.

1.1 Induction of labour with oxytocin

Induction of labour (IOL) with oxytocin is the artificial initiation of uterine contractions which leads to progressive effacement and dilatation of the uterine cervix, and descent of the fetus. While IOL with oxytocin alone is indicated in women with ruptured membranes, it is not recommended in women with intact membranes [6]. The objective of this intervention is to give the minimum effective dose until optimal myometrial contractions are achieved. Oxytocin can be added to normal saline for infusion, but large volumes of oxytocin infusion must not be administered due to the risk of hyponatremia. Oxytocin infusion has a half-life of 30 minutes, and by 40 minutes achieves steady-state levels. Therefore, oxytocin dose is increased at intervals of about 40 minutes [7]. There is no evidence to show whether a low-dose or a shigh-dose oxytocin regimens are optimal, and this compounds the confusion in clinical practice [5].

Oxytocin has very variable effects in terms of uterine contractions and fetal hypoxia in different subjects, which results in unpredictability of response in clinical setting. During increments, a close monitoring of labour progress becomes essential, anticipating that adverse effects of oxytocin on uterine activity and the fetus are exclusively dose-related [5]. The attention must be on uterine contractions and the fetus rather than the dosage of oxytocin.

Synthetic oxytocin is frequently used for IOL [5]. But its use has not been standardised and reported to be used improperly. Desirable effects on the brain that are demonstrated with physiologic oxytocin are absent with exogenous oxytocin infusions. It is advised that a protocol and algorithms with oxytocin, needs to be followed [5]. It is recommended by various authors to have an interval of 40 minutes between increments while observing the uterine contractions and the fetus closely. The use of oxytocin needs continuous supervision [5].

1.2 Oxytocics (also known as uterotonics)

These are agents that stimulate the myometrium or promote uterine contractions and hence increase the tone of the uterus. Uterotonics are used to induce or augment labour to stimulate delivery of the placenta and to prevent or treat PPH. Common uterotonic agents are synthetic oxytocin, synthetic oxytocin analogue carbetocin, methylergometrine, carboprost and misoprostol.

1.3 Postpartum haemorrhage

PPH is the leading cause of maternal morbidity and mortality globally, with atony of the uterus responsible for up to 80% of cases, thereby being the single most common cause [8, 9]. As is clear from **Figure 1**, retraction of uterine muscle fibres clamps the blood vessels and helps to stop bleeding postpartum. Therefore, uterine atony will result in clinical PPH. **Figure 2** shows why it is normal to expect bleeding postpartum, as placental site has a large surface area with cut blood vessels.

Conventionally, PPH is defined as blood loss of at least 500 ml after vaginal delivery and blood loss of >1000 ml after Caesarean section [10]. The American College of Obstetricians and Gynaecologists (ACOG) revitalise initiative defined PPH as cumulative blood loss of >1000 ml (irrespective of the route of delivery) or

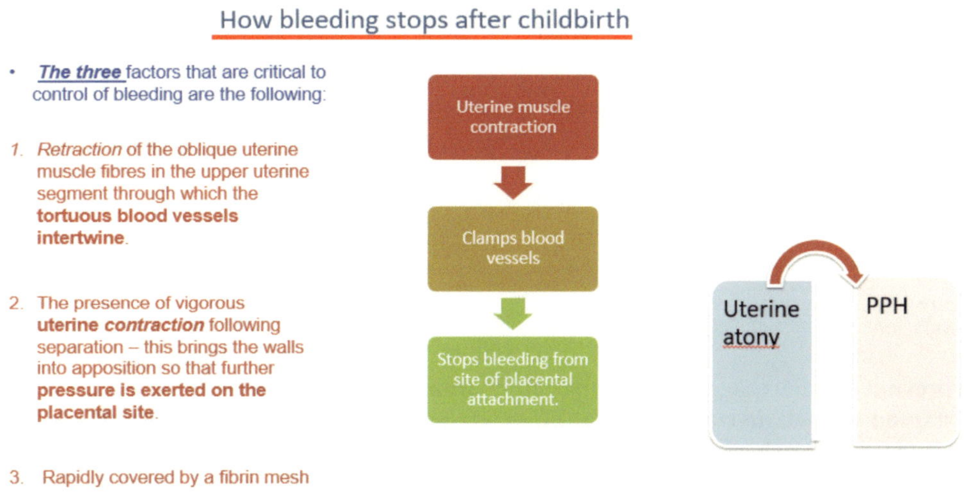

Figure 1.
Uterine contractions are essential to control bleeding after childbirth.

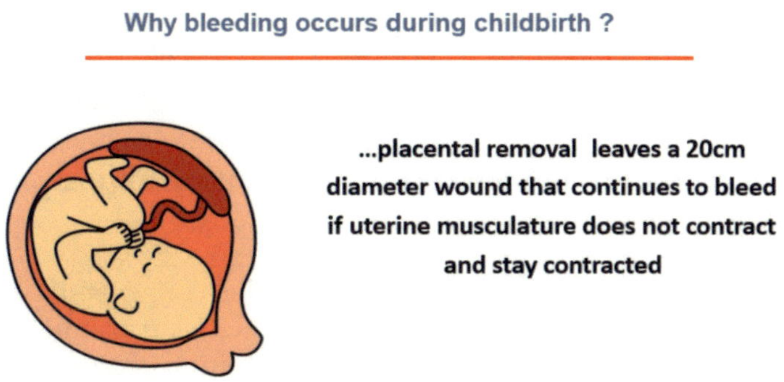

Figure 2.
Placental separation results in large wound that bleeds after childbirth.

blood loss accompanied by signs or symptoms of hypovolemia within 24 hours after the birth process.

It contributes greatly to significant maternal morbidity, long-term disability, and many other severe maternal conditions, which involve significant blood loss. These include anaemia, cardiac failure, and sepsis.

Carbetocin is a drug that precludes for additional uterotonic drugs in women at increased risk of PPH. Better trials are underway to evaluate carbetocin in preventing PPH in high-risk women [11].

1.4 Therapies for PPH

Oxytocin is the current standard drug for the prevention of PPH. Oxytocin availability in developing nations is limited by the requirement for temperature-regulated storage and administration by skilled nursing staff [12]. PPH prophylaxis with oxytocin fails commonly, necessitating the use of further oxytocin or other treatments to maintain stability [13]. The efficacy of uterotonics in causing uterine contractions

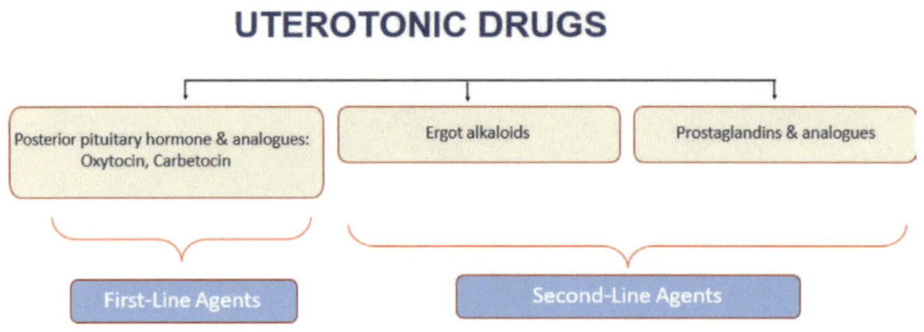

Figure 3.
Classification of uterotonic drugs.

to prevent haemorrhage can be impaired by improper storage. Where access to sustained cold-chain is unavailable, the efficacy of oxytocin cannot be ascertained as it is susceptible to heat [14, 15]. Being a short half-life drug, a continuous IV infusion is necessary for a sustained uterotonic effect. Boluses of oxytocin are associated with adverse effects like hypotension, nausea, vomiting, dysrhythmias, ST-T changes, pulmonary oedema and severe water intoxication with convulsions. Miscellaneous uterotonics include ergometrine/methylergometrine, and misoprostol. Ergometrine degrades on exposure to heat or light. Misoprostol degrades rapidly when exposed to moisture [16]. When degraded, the level of active ingredient is reduced, resulting in a loss of efficacy. The classication of uterotonic drugs is given in **Figure 3**.

2. Carbetocin: a therapy advance for the prevention of postpartum haemorrhage

Carbetocin is a novel drug, developed as long-acting congener of oxytocin.

It has shown similar pharmacologic traits as oxytocin, but around 10-times longer half-life than oxytocin. Carbetocin does not show variation in dose response, it lacks receptor desensitisation, and thus is an advance over oxytocin in this space [17].

As shown in **Figure 4**, the amino group and the disulfide bond, which were altered to create carbetocin, are indicated. The amino group was removed and the sulphur atom was replaced by a carba group.

Its long uterine activity is beneficial in the management of the third stage of labour. The side-effect profile of carbetocin is better than oxytocin and other uterotonics. Heat-stable carbetocin has demonstrated to maintain stability for 36 months at 30°C and 75% relative humidity [18].

2.1 Pharmacological properties of carbetocin

As a novel analogue of oxytocin, carbetocin has uterotonic activity by binding to oxytocin receptors on the myometrial cells. The main disadvantage of oxytocin is its short half-life (3–17 minutes) [19]. By modifying the oxytocin molecule, its half-life has been prolonged. Because of alterations, carbetocin has more pronounced pharmacological effects [20].

Carbetocin is a synthetic oxytocin analogue, with a potency of about one-tenth that of oxytocin [21]. Its plasma half-life is approximately 40 minutes, which is about 10 times longer than that of oxytocin [22]. It causes an increase in the intracellular concentration of calcium that promotes uterine contractility, via inositol phosphates signalling pathways [23].

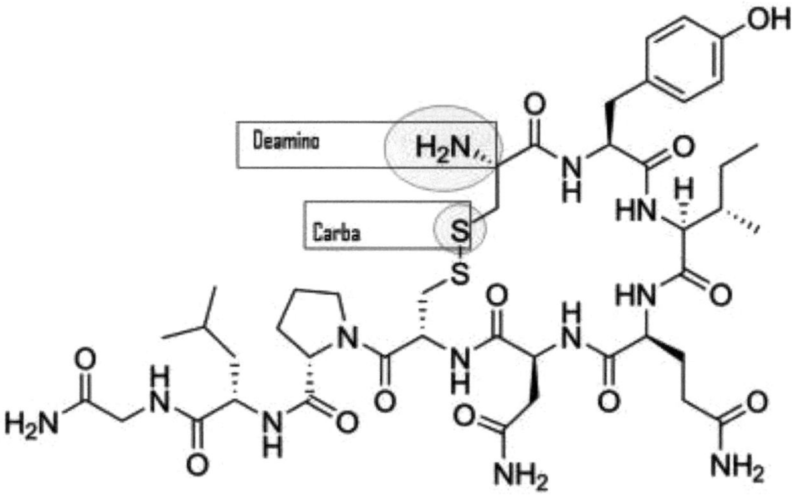

Figure 4.
Modification of oxytocin molecule.

The onset of action is rapid, with a firm contraction being obtained within 2 minutes in around 90% of patients. The duration of action of a single iv injection is about 1 hour, and approximately 2 hours when given as an IM injection. Carbetocin induces a prolonged uterine response, in parameters like amplitude and frequency of contractions.

Adverse reactions: The adverse reactions are the same as those with the use of oxytocin. Carbetocin was associated with nausea, abdominal pain, pruritus, flushing, and tremor (11%) [24].

2.2 Carbetocin to prevent haemorrhage after vaginal birth

Widmer et al. enrolled women across 23 sites in a randomised trial comparing IM injections of heat-stable carbetocin with oxytocin after vaginal birth [25]. The endpoints included the proportion of patients with blood loss of 500 ml or the additional uterotonics use, and patients with blood loss of 1000 ml. The authors concluded that carbetocin was non-inferior to oxytocin.

2.3 Carbetocin to prevent haemorrhage after caesarean delivery

This study enrolled women at risk of PPH after Caesarean section [26]. More than 1200 women were included. Around 750 received oxytocin first and around 480 received carbetocin first. It was demonstrated that compared with oxytocin, carbetocin reduced the need for uterotonics or interventions in high-risk patients.

2.4 Agents for in PPH prophylaxis meta-analysis

The uterotonics use during the third stage of labour for preventing PPH were compared with a control groups. The study demonstrated that ergometrine plus oxytocin combination, carbetocin, and misoprostol plus oxytocin combination were most efficacious in evaluated parameters. Carbetocin showed the best side-effect profile among the studied groups [27, 28].

The WHO did not include carbetocin in its 2012 guideline for PPH [29]. But in its update published in 2018, it has included carbetocin to the Essential Medicines List [30]. Carbetocin is recommended for the prevention of bleeding after all births when oxytocin is unavailable or its quality cannot be assured. Carbetocin remains effective at warm temperatures [31], while oxytocin has to bestored and transported at 2°C–8°C.

2.5 Carbetocin in emergency caesarean delivery

Patients were randomised to an iv injection of oxytocin or carbetocin after Caesarean delivery. Add-on uterotonics use was lower in the carbetocin group. Carbetocin was superior to oxytocin in evaluated parameters by 12% [32].

2.6 Dose and method of administration of carbetocin

- CARBETOCIN should be administered as a single dose only.

- Caesarean section: A single dose of 100 μg (1 ml) of CARBETOCIN injection should be administered iv as a bolus. CARBETOCIN can be administered either before or after delivery of the placenta.

- Vaginal delivery: A single dose of 100 μg (1 ml) of CARBETOCIN injection should be administered after delivery of the infant as an im or iv bolus injection slowly over 1 minute.

3. Miscellaneous uterotonics

3.1 Methyl-ergometrine

It is a vasoconstrictor and induces uterine contraction. It runs the risk of hypertension, coronary artery spasm and bronchospasm. Thus, it should not be used in concomitant cardiopulmonary diseases and pre-eclampsia. It may be used when the response to oxytocin is insufficient.

3.2 Carboprost

Carboprost is used as IM injection in 250 μg doses; can be repeated up to eight such doses (i.e. maximum 2 mg). Side effects are nausea, vomiting, diarrhoea, fever, bronchospasm and hypertension. It is used as a last resort. It should also be used with caution in active hepatic or cardiovascular disease.

3.3 Misoprostol

This prostaglandin analogue can replace oxytocin, if not available. It is used at a dose of 600 μg orally or sublingually in PPH. Diarrhoea, shivering, pyrexia and headache are some of the side effects. The drug is second-line agent when methylergometrine is contraindicated such as in preeclampsia.

4. Conclusion

Failure of prophylaxis with oxytocin in PPH (as demonstrated by the need for a rescue uterotonic) occurs commonly, necessitating the use of further oxytocin or other treatments to maintain haemodynamic stability. Uterotonics include ergometrine/methylergometrine & misoprostol, as shown in **Figure 3**. The major disadvantages of oxytocin are its short half-life (3–17 minutes) and its requirements for cold storage and transport. By modifying the oxytocin molecule, its half-life has been prolonged and its enzymatic degradation reduced. Carbetocin has more pronounced pharmacological effects. Its main advantage over oxytocin is a longer uterotonic activity, which obviates the need of a continuous infusion and has a standardised dosing of single injection recommendation, carbetocin can address the variations in dosing regimen as is with oxytocin.

The posology of carbetocin has tremendous benefit for the patient. Carbetocin selectively binds to oxytocin receptors present on the myometrium of the uterus, resulting in rhythmic contractions, increased frequency of existing contractions, and increased uterine tone. Another feature to note is that carbetocin selectively has a pronounced effect on the pregnant and immediate postpartum uterus.

Points to Remember

- Oxytocin is limited by the needs for cold storage and transport, and especially in low-income nations, this hampers its quality as there is a lack of such facilities.
- Because of its heat stability, carbetocin is an effective alternative in such a scenario.
- A more uniform dosing regimen of single injection, carbetocin can address the variations in the dosing (cf oxytocin).
- Carbetocin, is added to WHO Essential Medicines List for the prevention of excessive bleeding after childbirth.
- Carbetocin is a new paradigm in the prophylaxis of uterine atony.

Acknowledgements

Authors would like to express sincere gratitude to their mentors and esteemed teachers Dr. SC Chopra and Dr. (Mrs) Chanchal Gupta, former Heads, Department of Pharmacology of Dayanand Medical College, Ludhiana for providing the necessary support in getting the required information for this study. This chapter is dedicated to the faculty & staff of Department of Pharmacology of Dayanand Medical College, Ludhiana and the executives & staff of Uniza Healthcare LLP, Ahmedabad.

Competing interests

Author has declared that no competing interests exist.

Abbreviations

PPH	postpartum haemorrhage
IOL	induction of labour
ACOG	American College of Obstetricians and Gynaecologists
iv	intravenous
IM	intramuscular

Author details

Amit Bhalla[1]* and Sandeep Kaushal[2]

1 Uniza Healthcare LLP, Ahmedabad, India

2 Dayanand Medical College, Ludhiana, India

*Address all correspondence to: amitbhll@yahoo.com

References

[1] Kruse J. Oxytocin: pharmacology and clinical application. The Journal of Family Practice. 1986;**23**(5):473-479 PMID: 3534134

[2] Barbieri RL. Optimizing the use of oxytocin on labor and delivery [Internet]. 2020. Available from: https://www.mdedge.com/obgyn/article/234234/obstetrics/optimizing-use-oxytocin-labor-and-delivery [Accessed: 06 June 2023]

[3] Uvnäs-Moberg K, Ekström-Bergström A, Berg M, et al. Maternal plasma levels of oxytocin during physiological childbirth – a systematic review with implications for uterine contractions and central actions of oxytocin. BMC Pregnancy and Childbirth. 2019;**19**:285. DOI: 10.1186/s12884-019-2365-9 [Accessed: 24 May 2023]

[4] Fuchs AR, Fuchs F, Husslein P, Soloff MS. Oxytocin receptors in the human uterus during pregnancy and parturition. American Journal of Obstetrics and Gynecology. 1984;**150**(6):734-741

[5] Dasanayake DL, Goonewardene M. Use of oxytocin for induction of labour. Sri Lanka Journal of Obstetrics and Gynaecology. 2021;**42**(4):137-142

[6] World Health Organisation. WHO recommendations for induction of labour 2012, WHO, Geneva Switzerland

[7] Amico JA, Seitchik J, Robinson AG. Studies of oxytocin in plasma of women during hypocontractile labor. The Journal of Clinical Endocrinology and Metabolism. 1984;**58**(2):274-279

[8] Jones AJ, Federspiel JJ. Preventing postpartum hemorrhage (PPH) with combined therapy rather than oxytocin alone pharmacologic therapy. American Journal of Obstetrics & Gynecology MFM. 2022:100731

[9] Theunissen FJ, Chinery L, Pujar YV. Current research on carbetocin and implications for prevention of postpartum haemorrhage. Reproductive Health. 2018;**15**(1):55-59

[10] Su LL, Chong YS, Samuel M. Carbetocin for preventing postpartum haemorrhage. Cochrane Database of Systematic Reviews. 2012;**4**:CD005457

[11] Kalafat E, Gokce A, O'Brien P, Benlioglu C, Koc A, Karaaslan O, et al. Efficacy of carbetocin in the prevention of postpartum hemorrhage: a systematic review and Bayesian meta-analysis of randomized trials. The Journal of Maternal-Fetal & Neonatal Medicine. 2021;**34**(14):2303-2316

[12] Ayadi AM, Robinson N, Geller S, Miller S. Advances in the treatment of postpartum haemorrhage. Expert Review of Obstetrics & Gynecology. 2013;**8**(6):525-537

[13] Fahmy NG, Yousef HM, Zaki HV. Comparative study between effect of carbetocin and oxytocin on isoflurane-induced uterine hypotonia in twin pregnancy patients undergoing cesarean section. Egyptian Journal of Anaesthesia. 2016;**32**:117-121

[14] Hogerzeil HV, Gja W, de Goeje MJ. Stability of injectable oxytocics in tropical climates: WHO report; 1993. WHO/AP93.6. Available from: http://apps.who.int/iris/handle/10665/59411

[15] Mullany L, Newton S, Afari-Asiedu S, et al. Cumulative effects of heat

exposure and storage conditions of oxytocin-in-Uniject in rural Ghana: implications for scale up. Global Health: Science and Practice. 2014;**2**(3):285-294. DOI: 10.9745/GHSP-D-14-00043

[16] WHO Quality of misoprostol products. WHO Drug Information. 2016;**30**:35

[17] Gallos ID, Papadopoulou A, Man R, et al. Uterotonic agents for preventing postpartum haemorrhage: A network meta-analysis. Cochrane Database of Systematic Reviews. 2018;**12**:CD011689. DOI: 10.1002/14651858.CD011689.pub3. PMID: 30569545; PMCID: PMC6388086

[18] Malm M, Madsen I, Kjellström J. Development and stability of a heat-stable formulation of carbetocin for the prevention of postpartum haemorrhage for use in low and middle-income countries. Journal of Peptide Science. 2018;**24**(6):e3082

[19] American College of Physicians. Practice strategies for elective red blood cell transfusion. Annals of Internal Medicine. 1992;**116**:403-406

[20] Atke A, Vilhardt H. Uterotonic activity and myometrial receptor affinity of 1-deamino-1-carba-2-tyrosine(O-methyl)-oxytocin. Acta Endocrinologica. 1987;**115**:155-160

[21] Hunter DJS, Schulz P, Wassenaar W. Effect of carbetocin, a long-acting oxytocin analogue on the postpartum uterus. Clinical Pharmacology and Therapeutics. 1992;**52**:60-67

[22] van Dongen PWJ, Verbruggen MM, de Groot AN, van Roosmalen J, Sporken JM, Schulz M. Ascending dose tolerance study of intramuscular carbetocin administered after normal vaginal birth. European Journal of Obstetrics, Gynecology, and Reproductive Biology. 1998;**77**:181-187

[23] Engstrøm T, Barth T, Melin P, Vilhardt H. Oxytocin receptor binding and uterotonic activity of carbetocin and its metabolites following enzymatic degradation. European Journal of Pharmacology. 1998;**355**:203-210

[24] Pabal. Product information. Available at Pabal 100 micrograms in 1ml solution for injection - Summary of Product Characteristics (SmPC) - (EMC) (medicines.org.uk). [Last Accessed: 16 April 2022]

[25] Widmer M, Piaggio G, Nguyen TMH, Osoti A, Owa OO, Misra S, et al.; WHO CHAMPION Trial Group. Heat-stable carbetocin versus oxytocin to prevent hemorrhage after vaginal birth. The New England Journal of Medicine. 2018;379(8):743-752. DOI: 1056/NEJMoa1805489. Epub 2018 Jun 27. PMID: 29949473

[26] Tse KY, Yu FNY, Leung KY. Comparison of carbetocin and oxytocin infusions in reducing the requirement for additional uterotonics or procedures in women at increased risk of postpartum haemorrhage after Caesarean section. Hong Kong Medical Journal 2020;26(5):382-389. DOI: 10.12809/hkmj208683. Epub 2020 Oct 8. PMID: 33028723

[27] Gallos ID, Papadopoulou A, Man R, Athanasopoulos N, Tobias A, Price MJ, et al. Uterotonic agents for preventing postpartum haemorrhage: a network meta-analysis. Cochrane Database of Systematic Reviews. 19 Dec 2018;**12**(12):CD011689. DOI: 10.1002/14651858.CD011689.pub3. PMID: 30569545; PMCID: PMC6388086

[28] World Health Organization. WHO recommendations on prevention and

treatment of postpartum haemorrhage. Geneva: World Health Organization; 2012

[29] WHO. WHO recommendations: Uterotonics for the prevention of postpartum haemorrhage. Available from: https://apps.who.int/iris/bitstream/handle/10665/277276/9789241550420-eng.pdf?ua=1. [Last Accessed: April 2022]

[30] Malm M et al. Development and stability of a heat-stable formulation of carbetocin for the prevention of postpartum haemorrhage for use in low and middle-income countries. Journal of Peptide Science. 2018:e3082

[31] Al Zubaidi S, Alhaidari T. Heat stable carbetocin vs. oxytocin for the prevention of post-partum haemorrhage in emergency caesarean delivery: a randomized controlled trial. Journal of Perinatal Medicine. 2022;**50**(2):150-156. DOI: 10.1515/jpm-2021-0206

[32] Theunissen FJ, Chinery L, Pujar YV. Current research on carbetocin and implications for prevention of postpartum haemorrhage. Reproductive Health. 2018;**15**(1):55-59

Chapter 7

Social and Behavior Change Communication Framework

Olih Solihin, Yasundari Yasundari, Ahmad Zakki Abdullah, Kurniati Devi Purnamasari, Maulana Irfan and Yuni Mogot

Abstract

The Social and Behavior Change Communication (SBCC) framework is an approach used to influence and improve public health behaviors. This framework integrates principles of communication theory and social psychology to create more effective messages for behavior change. There are four stages in the SBCC framework: First, situational analysis involves collecting and analyzing data about the health behaviors that need to be changed. Second, planning and strategy is a continuation of the situational analysis, where the next step is to plan and select the appropriate communication strategy to achieve the desired goals. This strategy may involve delivering messages through mass media, information campaigns, or individual interventions. Third, message and communication material development involves developing relevant and appealing messages and communication materials for the public. Messages should be designed while considering social and cultural factors, language used, and media preferences used by the public. Fourth, evaluation is used to assess the effectiveness of messages and strategies used. Evaluation can be done by measuring changes in health behavior, public awareness of specific health issues, and factors that influence behavior. In health communication, the SBCC framework can help to increase public awareness of health issues, motivate them to change unhealthy behaviors, and encourage healthier behavior.

Keywords: social, behavior change, communication framework, health communication, situational analysis, stunting

1. Introduction

The problem of stunting persists in the world and poses a significant obstacle in attempts to attain optimal child welfare and development [1, 2]. Stunting, also known as stunted growth, happens when a child does not reach the physical growth that should occur at a specific age, resulting in him being shorter than the average child his age [3].

Stunting is common in many countries, especially underdeveloped countries and Indonesia and is an indicator of unresolved health and nutrition issues [4]. Stunting has substantial long-term consequences, including lower cognitive ability and intellect, increased risk of chronic disease in maturity, and diminished children's future potential for personal achievement and prosperity [5].

Although stunting rates have decreased in some nations in recent years, the obstacles that remain in combating stunting are complicated and varied. The main reasons of high stunting rates are malnutrition, insufficient food, restricted access to health services and clean water, poor sanitation, poverty, and socioeconomic inequality [4]. Stunting is also linked to poor feeding habits, such as not providing newborns with exclusive breastfeeding, not offering healthy supplementary foods after six months of age, and a lack of diversity and quality of food in a child's diet. Diet and child care are also influenced by social, cultural, and economic factors, which might increase the risk of stunting [6].

Overcoming the stunting problem is a complex endeavor that necessitates a multi-sectoral approach, institutional collaboration, and a strong commitment from the government, civil society organizations, the commercial sector, and society as a whole. Stunting efforts must be supported by policies that emphasize balanced nutrition, accessible and high-quality health care, and changes in people's behavior toward healthy eating practices.

Stunting is still a pressing issue in human development and world health. By recognizing the importance of addressing stunting and raising awareness of its long-term consequences, it is hoped that a stronger commitment and action may be taken to provide children with the protection and assistance they require to grow and develop properly [6].

From the standpoint of communication studies, one solution to addressing stunting is to present a holistic and integrated approach, including the use of the Behavior Change Communication Framework, which is a strategy that combines aspects of communication and behavior change to achieve the desired social change goals. Social Behavior Change Communication Framework is an excellent strategy for changing people's ideas, attitudes, and behavior toward balanced nutrition, good feeding, and optimal infant care in the context of combating stunting.

Indonesia has a distinct cultural, socioeconomic, and geographical variety. As a result, the Social Behavior and Change Communication (SBCC) that is adopted must take these contextual distinctions into account and build a communication strategy that is tailored to the peculiarities of the Indonesian people. The SBCC strategy is intended to raise awareness of the need of balanced nutrition and stunting prevention, as well as to encourage people to adopt behaviors that promote optimal growth and development. SBCC works with a variety of partners in Indonesia to combat stunting, including the government, health institutions, civil society, and the commercial sector. Collaboration among all of these stakeholders is critical in order to maximize efforts to combat stunting through an effective and coordinated communication approach [7].

In this chapter, we will examine further the implementation of SBCC in dealing with stunting in Indonesia. This approach is expected to provide better insight into appropriate communication strategies, effective communication channels, and stronger community involvement in stunting prevention efforts. The results of this research are expected to be a reference for policymakers and practitioners in designing and implementing effective and sustainable stunting prevention programs in Indonesia.

DOI: http://dx.doi.org/10.5772/intechopen.112744

2. Stunting cases in Indonesia

In 2020, there will be 149.2 million stunted, 45.4 million wasting, and 38.9 million overweight children worldwide. Except for Africa, all areas are seeing a decrease in the number of children with stunting. Asia as a whole is home to more than three-quarters of all children with severe wasting, and more than half of all children affected by wasting live there. At the national level, the stunting target is where countries are making the most success, with over two-thirds of them observing at least some improvement. In contrast, the majority of nations have either made little progress or worsened when it comes to the obesity [8].

As shown in **Figure 1**, globally, there has been improvement in the rates of chronic undernourishment and linear growth stunting in children under the age of five; however, these rates are still high in many areas. The growth and development of children may be enhanced by policies, programs, and interventions that support maternal and child health and nutrition. Data from the Indonesian Ministry of Health, one in three of the country's nine million children were stunted in 2018 [9, 10]. The Delivery of Additional Food program is then used nationwide to carry out the government's objective to combat stunting. This program delivers healthy biscuits to toddlers and pregnant women, with a special emphasis on undernourished toddlers.

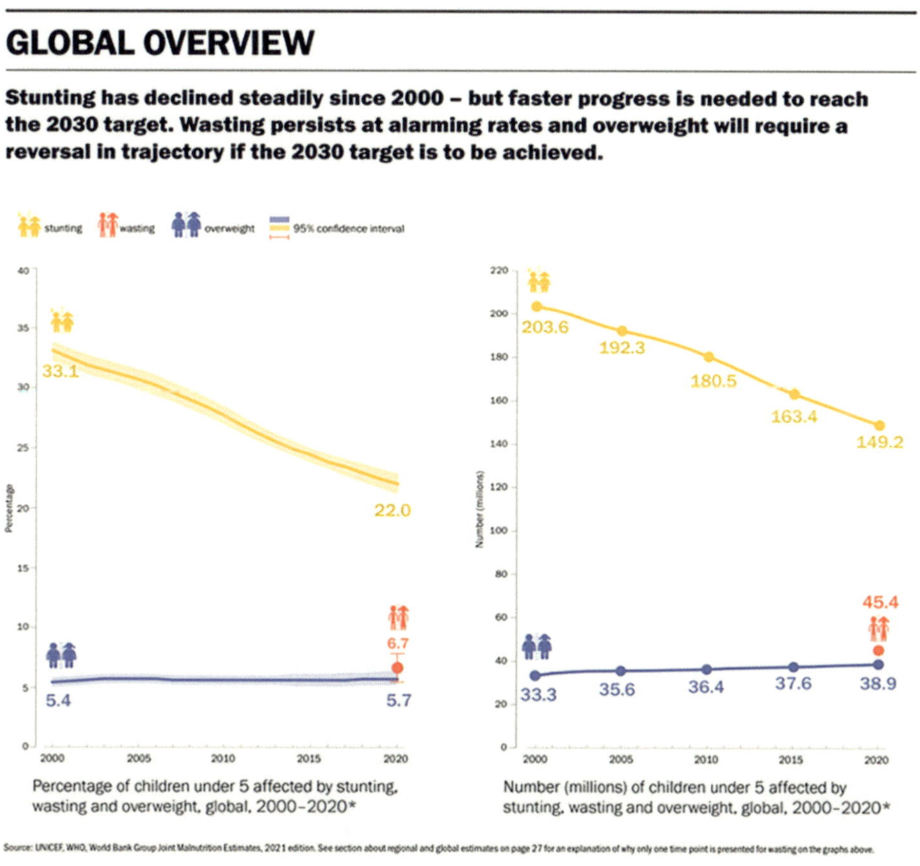

Figure 1.
Global overview of stunting [9].

In underdeveloped nations, stunting in toddlers is closely linked to hunger because chronic malnutrition affects the prevalence of stunting. Height and body weight can be used to assess growth and development, which can have an impact on both physical and psychological growth and development in toddlers [11]. To prevent stunting in children, dietary supplementation is, therefore, crucial during the toddler years.

By 2025, Indonesia aims to achieve its Sustainable Development Goal of a 40% reduction in stunting. Analyzing the dietary intake of children under the age of five hence requires more investigation. A critical time for a child's growth and development is when they are newborns and young infants to toddlers. The growth of the child's body and mind will be stunted by stunting situations [12, 13]. Stunting is typically brought on by children's inadequate nutritional intake, which is linked to diet and infection [14, 15]. One of the research-based risk factors for stunting indicates that environmental sanitation outcomes, such as the accessibility of latrines and the quality of drinking water, carry a higher risk [16, 17]. Lead exposure is linked to a number of nutritional deficiencies, which ultimately stunts children's neurodevelopment [18]. Children who consume contaminated food and are exposed to environmental toxins like metals are, especially susceptible to poor nutrition absorption. Stunted growth is one common sign of a nutritional deficiency and happens when a child's height (or length) for that age falls below the 5th percentile [17].

The previous research showed that stunted conditions (22%) with a z score of 3 to 2 standard deviations (SD), followed by severely stunted circumstances (8%) with a z score of 3 SD, dominated the evaluation of the stunting status of children under five in 2021. The study showed that the condition of stunting was not caused by environmental causes. However, it was thought that drinking boiled water increased the risk of stunting. Additionally, our data demonstrated a correlation between the state of stunting and exclusive breastfeeding, which supplied a protective factor for stunting in toddlers [19].

Ambitious international nutrition goals have been established during the current Sustainable Development Goals era. The World Health Assembly set the goal of reducing stunting in children under the age of five by 40% by 2025 as one of its objectives. According to the World Health Organization, stunting is the decreased growth and development of children as a result of inadequate nutrition, frequent infections, and insufficient psychosocial stimulation. If a child's height for age is less than −2 standard deviations from the WHO Child Growth reference, they are considered stunted. Though the overall number of stunted children has reduced, there are still 150 million of them in the globe under the age of five. This figure represents a tiny portion of the kids whose linear growth is being slowed down for a variety of reasons. The main focus is on supporting breastfeeding, complementary feeding, and anti-infection measures for kids. All of these are crucial for a child's health and survival, yet they have little to no impact on stunting. Adolescent females' health and nutrition must be promoted through action. Often, the first pregnancy occurs too early.

As shown in **Figure 2**, The Sustainable Development Goals Report 2022 provides a global overview of progress on implementation of the Agenda for Sustainable Development, using the latest available data and estimates. Prenatal nutrition is required in cases of food insecurity, along with other prenatal or early pregnancy interventions. "Reducing the burden of stunting requires a paradigm shift from interventions focusing solely on children and infants to those that reach mothers and families and improve their living environment and nutrition," the authors of the Danaei research write in their conclusion.

Indonesia's COVID-19 epidemic has had an effect on a number of areas, including the economy, education, and other facets of communal life, such as health issues.

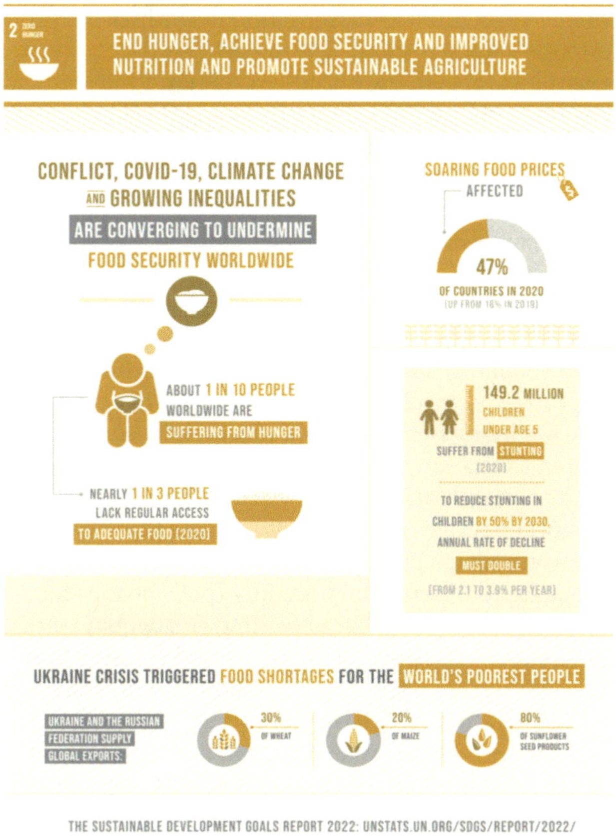

Figure 2.
The sustainable development goals report 2022 [20].

Despite being quite a heavy burden on the health sector, the government's various efforts to foresee the effects of the COVID-19 pandemic, particularly for vulnerable populations like pregnant women and young children, have produced quite encouraging results as the issue of stunting in Indonesia has decreased over the past two years. This evaluation of toddler nutrition status is also strongly related to the key objectives of the Healthy Indonesia Program in the 2020–2024 National Medium-Term Development Plan (NMTDP), namely enhancing the nutritional and health status of Indonesian children.

The national stunting rate has reduced by 1.6% every year, from 27.7% in 2019 to 24.4% in 2021, according to the findings of the 2021 SSGI [21]. Only 5 of the 34 provinces show a rise from 2019 while the majority of them show a fall. This demonstrates that the government's efforts to reduce stunting in Indonesia have shown some promising results. The Health Research and Development Agency of the Ministry of Health's SSGI 2021 study not only gives an overview of toddlers' nutritional status, but it can also be used as a tool for tracking and assessing the success of sensitive interventions and specific intervention indicators that have been implemented at the national and district/city levels starting in 2019 and lasting until 2024. At the moment, stunting is more common in Indonesia than in Myanmar (35%), although it is still more common than in Vietnam (23%), Malaysia (17%), Thailand (16%), and Singapore (4%).

Over the previous 10 years, Indonesia has maintained a high frequency of child stunting, which currently stands at about 37% nationwide. Uncertainty exists over the alignment of Indonesia's current child stunting prevention strategies with the available scientific data. We examine the available literature using the World Health Organization conceptual framework on child stunting to determine what has been examined, what can be said about the causes of child stunting in Indonesia, and where data gaps still exist [1]. Consistent evidence indicates that factors contributing to child stunting in Indonesia include nonexclusive breastfeeding for the first six months, low household socioeconomic position, early birth, short birth length, and low maternal height and education.

Figure 3 shows that the prevalence of stunting in children aged 0–59 months who live in homes with unimproved restrooms and untreated water are also more at risk throughout one decade. Child stunting has frequently been linked to neighborhood and socioeconomic problems, specifically a lack of access to health care and living in rural areas. There are not many published research on how education, society, culture, food production, water, sanitation, and the environment affect child growth. This thorough review of the research on the factors that influence child stunting in Indonesia identifies at-risk groups, effective interventions, and areas in which additional study is required to close knowledge gaps. According to the WHO framework, household and family factors, inadequate complementary feeding (poor quality foods, inadequate practices, and food and water safety), breastfeeding (inadequate practices), and infection (clinical and subclinical infection) are the main factors that contribute to child stunting. It classifies comparable contextual variables into the following subelements: political economics; health and health care; education; society and culture; agricultural and food systems; and water, sanitation, and environment. It also classifies contextual components into the general category of community and societal factors.

Stunting-risk family-based interventions are developed as a result of the accelerated program to reduce stunting, with an emphasis on preparing for parenthood, ensuring a healthy diet, improving parenting, expanding access to and improving the quality of health services, and expanding access to drinking water and sanitation. Monitoring and evaluation, the fifth pillar of the National Strategy for Stunting, is seen as critical and crucial in the quest to understand the effects of interventions on stunting prevention and control. This is anticipated to help lessen the issue of stunting in Indonesia in general and in priority districts/cities in particular [21]. Although the prevalence rate is currently below 20% in some places, it still falls short of the 14% target set by the NMTDP for 2024. Even though 14% has been met, stunting still exists in Indonesia. The next goal is to lower the stunting rate to a low category or below 2.5%.

Figure 3.
Prevalence of stunting (%) in children 0–59 months by district in 2013.

3. The role of social behavior and change communication

SBCC (Social and Behavior Change Communication) is a method of changing behavior and social practices in order to improve society's health, well-being, and progress. To achieve the desired change, SBCC integrates components of communication, behavior modification, and community participation [22].

The major purpose of SBCC is to positively influence and modify individual and group behavior. This is accomplished through the delivery of clear, relevant, and convincing messages, as well as the creation of an atmosphere conducive to behavior change. SBCC is a two-way process in which message recipients and message senders communicate continuously [23].

Social and Behavior Change Communication plays a critical role in dealing with stunting through lobbying, social mobilization, and education to achieve the behavior change required for preventing and overcoming stunting. The following is a detailed explanation of the SBCC's role in this context:

Advocacy: SBCC advocacy focuses on influencing policy, funding allocation, and support from key stakeholders. SBCC may express essential messages about stunting to policymakers, government agencies, and donor agencies through advocacy, with the goal of increasing awareness of the malnutrition problem, pushing necessary policy changes, and increasing financing for stunting treatment [24].

SBCC advocacy also includes working with civil society organizations, advocacy groups, and other stakeholders to raise awareness and demand for stunting. SBCC can use this strategy to promote policy changes, advocate for proper financial allocation, and create frameworks that support effective nutrition programs.

- Social mobilization entails actively involving the community in changing behavior and practices that promote stunting prevention. To engage and motivate the community to take steps that lower the risk of stunting, SBCC employs a communication approach that includes public campaigns, lectures, group discussions, and other participation activities. SBCC can raise public knowledge about the need of balanced nutrition and effective feeding habits, such as exclusive breastfeeding, timely supplemental feeding, and infant growth monitoring, through social mobilization. SBCC can also promote active community participation in existing nutrition programs, facilitate groups of mothers and families to share information and experiences, and gain community support for good change.

- Education is a crucial component of SBCC in combating stunting. Through education, SBCC provides people, families, and communities with accurate and relevant information on nutrition, child growth and development, and stunting prevention techniques. SBCC education includes learning about balanced nutrition, the importance of exclusive breastfeeding, the introduction of appropriate solid meals, and the importance of paying special attention to children's nutrition. Furthermore, teaching entails imparting motivational and inspirational messages about the positive effects of healthy dietary habits and the harmful repercussions of stunting. SBCC education also focuses on developing practical skills such as cooking, food selection, and how to evaluate children's growth. SBCC can help communities understand the importance of behavior change and provide the tools and methods needed to implement good nutrition habits in everyday life by using an interactive and participatory educational approach. SBCC teaching also helps to dispel myths and incorrect assumptions about nutrition and stunting. SBCC can influence people's attitudes and ideas through correct information and an effective communication technique, allowing them to adopt healthier nutritional practices and prevent stunting.

4. SBCC intervention in stunting management

SBCC interventions (Social and Behavior Change Communication) are communication methods and activities that try to change people's behavior, attitudes, and practices in order to improve their health, well-being, and quality of life [22]. SBCC interventions aim to address social and health issues by disseminating information, changing views, raising awareness, and pushing individuals and communities to adopt healthier habits. SBCC intervention entails employing an effective communication strategy, either verbal or nonverbal messages, to achieve the targeted behavior change [25]. This is accomplished by comprehending and respecting the intended target group's requirements, beliefs, culture, and social context.

SBCC interventions aim to promote long-term behavior change by affecting individuals' and society's knowledge, attitudes, norms, beliefs, and skills. This can involve health promotion, nutrition improvement, contraceptive use, sanitation behavior modification, disease control behavior, and other activities. SBCC intervention also entails determining the best communication channels, such as mainstream media, social media, formal and non-formal education, individual counseling, discussion groups, and mass communication campaigns. SBCC interventions use an integrated strategy to achieve a significant impact in enhancing people's welfare and addressing many social and health challenges [26].

Interventions in tackling stunting involve a series of communication efforts aimed at changing behavior and practices related to nutrition and child care. SBCC interventions in handling stunting, especially in Indonesia, as shown in **Figure 4**.

According to **Figure 4**, the components of SBCC intervention in handling stunting from the smallest part is from the individual level and the largest part is based on political system. All of the levels are explained as follows:

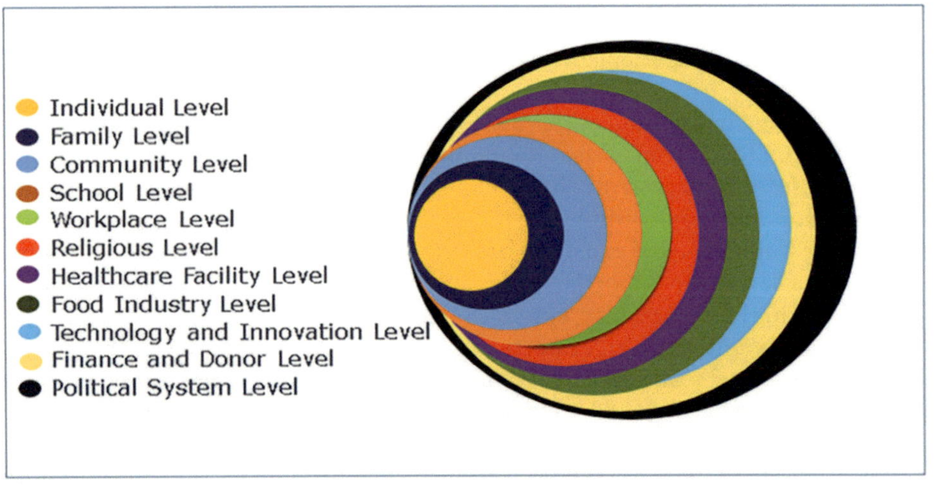

Figure 4.
SBCC intervention in handling stunting.

4.1 Individual level

At the individual level, SBCC interventions in handling stunting aim to change the behavior and practices of individuals and family members that have a direct impact on nutrition and child care [27]. The following describes SBCC interventions at the individual level:

- Counseling and Education: Through counseling and education, individuals, especially pregnant women, nursing mothers, and other family members, are given appropriate information about the importance of good nutrition and child care. This counseling can be carried out by health workers or competent nutrition service providers. They provide support and one-on-one guidance to individuals, helping them understand healthy practices that are important in stunting management.

- Personal Approach: Through a personal approach, SBCC interventions at the individual level can be tailored to the unique needs and situation of the individual or family. This approach includes intensive and ongoing communication between service providers and individuals, allowing individuals to share their problems, questions, and concerns face-to-face. With this approach, service providers can provide more personalized support and assist individuals in adopting appropriate nutrition and child care practices.

- Child Growth Monitoring: SBCC interventions at the individual level also involve the importance of regular monitoring of child growth. Individuals are given an understanding of how to monitor children's growth, such as measuring weight, height, and head circumference. This information assists the individual in quickly identifying a child's growth problem and seeking medical attention if needed.

- Education on Exclusive Breastfeeding: One of the main focuses at the individual level is to provide education about the importance of exclusive breastfeeding during the first six months of a child's life. Individuals are given information about the benefits of breastfeeding for the growth and development of children, as well as the right way to give exclusive breastfeeding.

- Healthy and Balanced Feeding Practices: SBCC interventions at the individual level also aim to increase understanding and adoption of healthy and balanced feeding practices in children. Individuals are given information about nutritious food, the right portion, the frequency of feeding, and the introduction of appropriate complementary foods when the child reaches the appropriate age.

Through SBCC interventions at the individual level, it is hoped that individuals and family members can obtain accurate knowledge, understand healthy practices, and implement the necessary behavioral changes to ensure optimal nutrition and care for children.

4.2 Family level

SBCC intervention at the family level aims to give family members with information, education, and assistance in the prevention of stunting. Families can grasp the importance of balanced nutrition and excellent feeding practices for children's growth and development through effective communication. Messages were distributed to raise awareness of the need of nutrition in the first 1000 days of life, from conception to the age of two.

At the family level, SBCC interventions can include nutritional advice, healthy eating behaviors, exclusive breastfeeding, the introduction of suitable solid meals, and effective child care practices. Families are also encouraged to frequently evaluate their children's growth, adhere to immunization regimens, and deal with nutritional issues on a proactive basis.

SBCC interventions at the family level also include psychosocial support and the enhancement of parental abilities in coping with child nutrition issues. Interpersonal contact between parents and children is improved in order to foster a positive atmosphere that is responsive to the needs of children. It is envisaged that the SBCC intervention at the family level will result in behavioral changes such as increased frequency of appropriate and balanced nutrition feeding, exclusive breastfeeding, and increased attention to children's health and nutrition. Stunting prevention programs can be more effective and long-lasting with active family participation.

4.3 Community level

SBCC interventions at the community level include a variety of activities and communication tactics geared at specific groups of people with significant social contacts and cultural ties. The primary purpose is to raise awareness, change attitudes, and support activities that can help avoid stunting.

Interventions at the community level can include health education, group discussions, awareness campaigns, and participatory activities engaging community members in the context of dealing with stunting. The messages provided in this intervention will be tailored to the cultural context, language, and local requirements of the community so that they are properly received.

Furthermore, SBCC interventions at the community level might include the active participation of community leaders or community leaders with influence within their circles. These individuals can serve as change agents by informing community members about the importance of balanced diet and stunting prevention.

Collaborative efforts with numerous local stakeholders, such as community organizations, NGOs, health personnel, and educational institutions, are also critical during community interventions. This collaboration has the potential to improve the implementation of SBCC treatments, broaden the breadth of messaging, and establish a community atmosphere that encourages desirable behavior change.

SBCC interventions at the community level are critical in building an environment that promotes stunting prevention and long-term behavior modification. It is intended that through enlisting the active participation and involvement of community people, common awareness, social support, and good norms linked to balanced nutrition and optimal infant care will be established in order to avoid stunting.

4.4 School level

SBCC interventions at the school level entail efforts to communicate relevant messages about healthy nutrition and stunting prevention to kids, teachers, and parents in the school setting. Here are some examples of SBCC interventions that can be implemented at the school level:

- Health Education: Health education initiatives in schools can focus on healthy diet and stunting prevention. This could be a lecture, seminar, or workshop that informs youngsters on the need of balanced diet and healthy eating practices. These messages can also emphasize the significance of exclusive breastfeeding, the proper introduction to solid meals, and other preventive actions.

- School Campaigns and Activities: Schools can initiate programs focusing on good nutrition and stunting prevention. They can, for example, arrange poster contests, theater performances, or other creative activities to improve students' knowledge of the necessity of nutritious food and healthy lifestyle choices. These types of activities can include kids, teachers, and even parents to encourage active engagement and good behavior modification.

- Healthy School Canteens: Schools can develop healthy canteen policies to meet the nutritional needs of their kids. This involves providing nutritious food and beverages, supporting local and organic foods, and avoiding foods rich in sugar and saturated fat. Messages regarding the benefits of eating nutritious foods and developing excellent eating habits can also be displayed in the canteen.

- Parental Involvement: By hosting specific meetings or communication sessions, schools can involve parents in SBCC interventions. On this occasion, parents can be given information about healthy nutrition and stunting prevention, as well as assistance and guidance on proper feeding methods and the importance of living a healthy lifestyle. Stunting prevention advocacy can also benefit from increased collaboration between schools and parents.

SBCC's school-based programs aim to provide an educational environment that promotes health and stunting prevention. This intervention can give knowledge, impact attitudes, and support good behavior changes in terms of healthy nutrition and stunting prevention in the school setting by including kids, teachers, and parents.

4.5 Workplace level

SBCC interventions in the workplace include efforts to raise workers' understanding and practice of health, as well as efforts to build work environments that promote health and stunting prevention. Here are some examples of SBCC initiatives at this level:

- Health Education: Workplace health education programs can be held to enlighten employees about the importance of balanced diet and stunting prevention. Trainings and workshops on issues such as exclusive breastfeeding, healthy feeding practices, and the necessity of appropriate child care can be organized. These messages can be delivered in the workplace through presentations, printed materials, or movies.

- Health Promotion: SBCC activities may also include workplace health promotion initiatives. Posters, pamphlets, or messages on internal media, for example, can be used to remind employees about the need of balanced diet and stunting prevention methods. Workplace arrangements that promote access to nutritious food and provide separate areas for breastfeeding or storing breast milk might also be encouraged.

- Employee Involvement: Involving employees in the formulation and implementation of health programs is an important part of SBCC interventions at the workplace level. Workers can contribute to program design and provide input on their needs and preferences using the participatory approach. Setting up workplace health committees or designating health leaders among employees could be examples of this.

- Community Support: The workplace can serve as a community where employees interact socially and support one another. SBCC interventions could take advantage of this by developing support groups or community-based activities in the workplace that promote good nutrition and stunting prevention. Workers can share information, experiences, and incentive to maintain excellent health and nutrition in this setting.

- Monitoring and Feedback: SBCC interventions at the workplace level may also include health-related monitoring and feedback systems. Regular nutritional status assessments or child development monitoring programs, for example, can be implemented to ensure personnel understand and follow desired health practices. Regular feedback to employees and management can encourage good behavior change.

4.6 Religious level

SBCC religious interventions involve collaboration with religious leaders and community leaders who play crucial roles in the community. The goal is to deliver messages about balanced nutrition and stunting prevention by employing societal religious values and religious beliefs. The SBCC intervention uses this strategy to promote positive and long-term behavior change in the community.

A variety of activities can be carried out as part of the SBCC intervention at the religious level, such as incorporating nutrition and stunting issues into religious lectures, recitations, or sermons in places of worship. Important messages regarding the necessity of balanced diet and stunting prevention methods can be transmitted by employing community-relevant allusions and holy texts.

Furthermore, small group conversations with religious and community leaders can be arranged to address the role of religion in dealing with stunting. Through open and participatory discourse, awareness, and greater understanding of religion's crucial role in supporting proper nutrition and preventing stunting can be established.

SBCC religious involvement may also include engagement with religious institutions such as religious-based religious groups, foundations, or social institutions. Through this collaboration, nutrition and stunting prevention health education programs can be linked into social activities such as social service activities, fundraisers, or support programs for persons in need.

SBCC intervention at the religious level is expected to influence behavior and rally societal support. This intervention can harness the power of religious principles in promoting healthy diets, sound nutritional practices, and the role of families and communities in stunting prevention by involving religious leaders and community leaders as change agents.

4.7 Healthcare facility level

Communication and techniques are focused at health professionals and health facility management at the Health Facility (Faskes) level [28]. The goal is to improve their understanding, abilities, and behavior in providing effective stunting management health services.

- Health Workers Training: Health workers in health facilities can obtain stunting management training and capacity building. This training may involve an understanding of stunting, risk factors, early detection, child growth monitoring, and optimal feeding practices. SBCC can also be used in this training to highlight the importance of their involvement in detecting and delivering early intervention for stunting.

- Monitoring and Feedback: SBCC can be used to improve child growth monitoring in health care institutions. Health practitioners can explain the importance of correctly measuring and recording children's growth by using clear and consistent messages. They can also give parents feedback on their child's development and make recommendations for nutritional improvements or interventions as needed.

- Educational materials and information: SBCC at the Faskes level can entail providing health staff and parents with suitable educational materials and information. This material could contain health care recommendations, pamphlets, posters, or booklets about balanced nutrition, proper feeding practices, and stunting prevention strategies. Health workers can help parents realize the importance of their participation in avoiding stunting by providing clear and easy-to-understand information.

- Counseling and Support: SBCC at the Faskes level can also involve parent and family counseling and support. Health professionals can advise on appropriate eating habits, breastfeeding, when to introduce solid meals, and proper feeding techniques. Furthermore, emotional and motivational assistance can be provided to ensure that families are actively participating in stunting prevention.

- SBCC intervention at the Health Facility level is critical for delivering quality health services and successful stunting prevention. Through effective communication and the correct methodology, health workers can become change agents in the fight against stunting at the individual and family levels.

4.8 Food industry level

The goal at the food industry level is to encourage food makers and retailers to produce, distribute, and promote nutritious food. These interventions include communications to food sector players such as manufacturers, retailers, and food companies [29].

One strategy for SBCC interventions in the food business is to communicate messages that inspire positive changes in food production and promotion practices. To raise knowledge about the need of balanced nutrition and its impact on stunting, communication initiatives might be established. These messages should include enhancing food product nutritional quality, reducing salt, sugar, and saturated fat, and increasing the availability and marketing of healthy foods, particularly for youngsters.

Food sector intervention could also entail the development of nutritional guidelines or standards for food goods produced and marketed in Indonesia. Working with health authorities and regulatory bodies to promote policies that support the adoption of better dietary guidelines is a part of this. SBCC interventions at this level could focus on the role of responsible advertising and promoting healthy food as a better choice in terms of promotion and marketing.

Furthermore, SBCC engagement in the food business might drive product innovation and the development of nutrient-rich functional meals. Effective communication can be utilized to raise the food industry's awareness of children's nutritional needs and encourage collaborative efforts to create better, stunting-prevention products. Because it can alter the availability and accessibility of healthy food in the market, SBCC involvement at the food industry level is a crucial step in attempts to combat stunting. It is thought that through modifying the food industry's behavior and methods, it will be possible to establish a food environment that is more supportive of child development and avoid stunting in Indonesia.

4.9 Technology and innovation level

At the level of innovation and technology, it entails using technology and innovation as a tool to communicate messages about health and behavior change to the general public. It seeks to promote comprehension, awareness, and good choice making [30].

SBCC interventions at the level of innovation and technology may involve employing mobile applications, text messages, digital platforms, or social media to deliver health information and education to parents or caregivers of children in the context of tackling stunting. Technology can also be utilized to provide reminders and assistance in the adoption of appropriate nutrition practices and the avoidance of stunting. Parents, for example, can receive nutrition, parenting, and good eating practices information via a mobile app.

This program can also serve as a reminder for feeding schedules and trips to medical institutions, as well as practical advice on appropriate food choices. As a result, technology can facilitate access to important information and encourage positive behavior change [31].

Furthermore, digital platforms like social media can be used to disseminate campaigns and information about healthy nutrition and stunting prevention. These messages can reach a wider audience through interesting and simple material, particularly among the younger population and young mothers who use social media. Online conversations, forums, or virtual support groups can also be developed to enable the exchange of information and experiences among parents or families who are worried about child nutrition or have stunting concerns.

SBCC interventions that are innovative and technological in nature have the ability to reach a larger audience and provide better accessibility to the community. However, it is critical to address the digital gap that may exist in specific areas or among certain categories of people. In this scenario, initiatives must be taken to guarantee that all sectors of society have access to and successfully utilize technology in order to combat stunting.

4.10 Finance and donor level

SBCC's interventions at the funding and donor levels aim to mobilize the support and resources needed to tackle stunting [32]. At this level, strategic communication and advocacy is used to engage financial institutions and donors in understanding stunting issues, the importance of preventive action, and the urgent need to address the problem of malnutrition among children in Indonesia.

SBCC interventions at the financial and donor levels entail delivering compelling messages and persuading them to participate in existing nutrition programs or offer more cash for stunting initiatives. These messages contain information regarding the long-term effects of stunting on child growth and development, as well as the consequences for human resource quality and overall country development [33].

Furthermore, interventions at the financing and donor levels include effective communication to stakeholders such as international financial institutions, donor organizations, and appropriate government agencies about nutrition issues. This is intended to motivate them to invest resources and take measures compatible with the goals of dealing with stunting.

SBCC interventions at the financing and donor levels also include advocacy, whether through meetings, presentations, or continuing discussion. The goal of this lobbying is to influence policies and budget allocations at the national and international levels so that stunting is prioritized and adequately funded.

To achieve the goals of SBCC interventions at the funding and donor levels, it is critical to collect robust data and evidence about the positive impact of nutrition interventions on society, as well as to provide regular reports and evaluations on the use of funds and the outcomes of the programs implemented. This will boost donor and financial institution trust and involvement in efforts to combat stunting in Indonesia.

4.11 Political and system level

SBCC interventions at the political and system levels are aimed at modifying policies, regulations, and service delivery mechanisms to support stunting prevention. At this level, the SBCC can lobby policymakers to boost stunting handling priorities and expenditures, develop regulatory frameworks for nutrition and child care, and ensure collaboration between agencies and associated sectors [34]. The goal is to establish an enabling environment in society for behavior change and sustainable practices. Some approaches that can be employed in SBCC interventions are as follows:

- Education and Information: Provide individuals, families, and communities with accurate and relevant information about healthy nutritional practices and effective child care. This can be accomplished through education, communication campaigns, flyers, and other forms of media.

- Counseling and Personal Approach: Enlisting the help of healthcare specialists to give nutrition counseling and care to people and families. Personal therapy might be provided to provide more intensive assistance and motivation.

- Public Relations Campaign: Using mass media such as television, radio, and social media to communicate vital nutrition and child care information to the larger population. To boost the appeal and impact of the message, this campaign might be supplemented with testimonials from prominent figures or celebrities.

- Using Local Culture: Using local culture to transmit SBCC messages. This can be accomplished through the use of traditional art, folklore, or local festivals that can draw attention and affect social standards in society.

- Policy Change and Advocacy: Advocating for policymakers to prioritize and promote stunting elimination. Sharing scientific evidence, conducting meetings and campaigns, and partnering with appropriate organizations and institutions to promote laws and regulations that support stunting reduction are all part of this effort.

- The goal of SBCC's stunting prevention intervention in Indonesia is to establish an environment that promotes healthy nutrition and child care practices, as well as to raise individual and community awareness, knowledge, and skills in dealing with stunting. This intervention, with the involvement of numerous stakeholders, is predicted to have a substantial impact on reducing stunting rates and enhancing the quality of life for children in Indonesia.

To achieve the goals outlined, the SBCC stunting strategy must be implemented through a staged method based on the SBCC model and planning principles [35]. This method is founded on evidence and a vision for a participative and long-term effort that incorporates all parties involved in dealing with stunting [36]. The steps that can be taken can be seen in **Table 1** below:

Stages	Description
Situation analysis	This stage involves an in-depth analysis of the situation related to the issue to be resolved. The purpose of the situational analysis is to understand the factors influencing behavior and practice regarding the issue, and to identify the target group to address. Situation analysis includes the collection of data and information, and the use of various research tools and methods to gain a comprehensive understanding of the context and characteristics of the target group.
Setting goals and targets	This stage involves setting clear and measurable goals and objectives for the SBCC program. These goals and objectives must be specific, measurable, achievable, relevant, and time-limited. In setting goals and objectives, various factors such as the level of awareness, knowledge, attitudes, and behavior that need to be changed are carefully considered.
Strategy design and development	This stage involves designing a strategy that will be used in the SBCC program. An effective strategy must be based on a deep understanding of the target group, context, and factors influencing behavior. Strategies can cover a variety of approaches such as media campaigns, individual counseling, training, discussion groups, etc. Strategy design also involves selecting appropriate messages, communication channels, and identifying suitable evaluation methods.
Material development and communication tools	This stage involves developing materials and communication tools that will be used in the SBCC program. Communication materials and tools must be adapted to the target group and the intended context. This includes creating messages that are attractive, clear, and appropriate to the needs of the target group, as well as selecting the right communication channels to reach the target group effectively.
Implementation	The implementation phase involves implementing the SBCC program according to a predetermined plan. This includes distribution of communication materials, delivery of messages, training, counseling, and various other activities that have been planned in the strategy. Implementation must be done with care, ensuring that the message is conveyed correctly, interaction with the target group occurs, and the necessary support is provided.
Monitoring and Evaluation	Monitor and evaluate the implemented SBCC program, to monitor progress, measure results, and make adjustments if necessary.

Table 1.
SBCC model and planning principles.

Author details

Olih Solihin[1*], Yasundari Yasundari[1], Ahmad Zakki Abdullah[2], Kurniati Devi Purnamasari[3], Maulana Irfan[4] and Yuni Mogot[1]

1 Indonesia Computer University, Bandung, Indonesia

2 Veteran National Development University of Jakarta, Jakarta, Indonesia

3 Universitas Galuh, Ciamis, Indonesia

4 Padjadjaran University, Sumedang, Indonesia

*Address all correspondence to: olih.solihin@email.unikom.ac.id

References

[1] Beal T, Tumilowicz A, Sutrisna A, Izwardy D, Neufeld LM. A review of child stunting determinants in Indonesia. Maternal & Child Nutrition. 2018;**14**(4):e12617

[2] Organization WH. Childhood stunting: Challenges and opportunities. In: Report of a Promoting Healthy Growth and Preventing Childhood Stunting Colloquium. Geneva: WHO; 2014

[3] Soofi SB, Ariff S, Khan GN, Habib A, Kureishy S, Ihtesham Y, et al. Effectiveness of unconditional cash transfers combined with lipid-based nutrient supplement and/or behavior change communication to prevent stunting among children in Pakistan: A cluster randomized controlled trial. The American Journal of Clinical Nutrition. 2022;**115**(2):492-502

[4] Dewey KG. Reducing child stunting: Moving forward on evaluating effectiveness of programs. The Journal of Nutrition. 2020;**150**(11):2843-2845

[5] Soofi SB, Khan GN, Ariff S, Rizvi A, Hussainyar MA, Garzon C, et al. Effectiveness of specialized nutritious foods and social and behavior change communication interventions to prevent stunting among children in Badakhshan, Afghanistan: Protocol for a quasi-experimental study. Methods Protocol. 2021;**4**(3):55

[6] Saleh A, Syahrul S, Hadju V, Andriani I, Restika I. Role of maternal in preventing stunting: A systematic review. Gaceta Sanitaria. 2021;**35**:S576-S582

[7] Yusuf M, Negara CK. Performance optimization of nursing management information system based on the fishing application (stunting monitoring information system) to increase the effectiveness of child nutritional health in primary health services. Journal of Technology. 2022;**1**(1)

[8] WHO. The UNICEF/WHO/WB Joint Child Malnutrition Estimates (JME) Group Released New Data for 2021. 2021

[9] Organization WH. Levels and trends in child malnutrition: UNICEF/WHO/The World Bank Group joint child malnutrition estimates: Key findings of the 2020 edition. In: Levels and Trends in Child Malnutrition: UNICEF/WHO/The World Bank Group Joint Child Malnutrition Estimates: Key Findings of the 2020 Edition. 2020

[10] Data P, Informasi Kemenkes RI. Situasi Balita Pendek (Stunting) di Indonesia. Buletin Jendela Data dan Informasi Kesehatan, Jakarta in Kementerian Kesehatan RI. 2018;**301**(5)

[11] Mitra M. Stunting problems and interventions to prevent stunting (a Literature Review). Jurnal Kesehatan Komunitas. 2015;**2**(6):254-261

[12] Berendsen MLT, Smits J, Netea MG, van der Ven A. Non-specific effects of vaccines and stunting: Timing may be essential. eBioMedicine. 2016;**8**:341-348

[13] Mukhopadhyay C, Vishwanath S, Eshwara VK, Shankaranarayana SA, Sagir A. Microbial quality of well water from rural and urban households in Karnataka, India: A cross-sectional study. Journal of Infection and Public Health. 2012;**5**(3):257-262

[14] Rodrigues JL, Batista BL, Nunes JA, Passos CJS, Barbosa F Jr. Evaluation of the use of human hair

for biomonitoring the deficiency of essential and exposure to toxic elements. Science of the Total Environment. 2008;**405**(1-3):370-376

[15] Yabe J, Nakayama SMM, Ikenaka Y, Yohannes YB, Bortey-Sam N, Kabalo AN, et al. Lead and cadmium excretion in feces and urine of children from polluted townships near a lead-zinc mine in Kabwe, Zambia. Chemosphere. 2018;**202**:48-55

[16] Bellinger DC. Very low lead exposures and children's neurodevelopment. Current Opinion in Pediatrics. 2008;**20**(2):172-177

[17] Campisi SC, Cherian AM, Bhutta ZA. World perspective on the epidemiology of stunting between 1990 and 2015. Hormone Research in Pædiatrics. 2017;**88**(1):70-78

[18] Handayati SP, Nasoetion A, Sukandar D. Konversi satuan ukuran rumah tangga ke dalam satuan berat (gram) pada beberapa jenis pangan sumber protein. Jurnal Gizi Dan Pangan. 2008;**3**(1):49-60

[19] Nurjazuli N, Budiyono B, Raharjo M, Wahyuningsih NE. Environmental Factors Related to Children Diagnosed with Stunting 3 Years Ago in Salatiga City, Central Java. Toxicologie Analytique et Clinique: Indonesia; 2023

[20] Department of Economic and Social Affairs Sustainable Development. The Sustainable Development Goals Report 2022. Goal 2 Infographic. 2022. Available from: https://unstats.un.org/sdgs/report/2022/

[21] Kemenkes RI. Penurunan Prevalensi Stunting Tahun 2021 sebagai modal menuju generasi emas Indonesia 2045. 2022

[22] McKee N, Becker-Benton A, Bockh E. Social and behavior change communication. 2014

[23] Egan KL, Wolfson M, Lukacena KM, Zelaya CM, McLeary MS, Helme DW. Developing a health communication campaign for disposal of unused opioid medications. Addictive Behaviors Reports. 2020;**12**:100291 Available from: https://www.sciencedirect.com/science/article/pii/S2352853220301061

[24] Núñez A, Madison M, Schiavo R, Elk R, Prigerson HG. Responding to healthcare disparities and challenges with access to care during COVID-19. Health Equity. 2020;**4**(1):117-128

[25] Almossawi H, Longacre CF, Pillay YG, Kak N, Chase C, Africa S, et al. A social and behavior change communication framework for addressing delays to appropriate TB care and treatment. Journal of Lung and Health Diseases. 2019;**2019**

[26] Solihin O, Muslim M, Sari IU. Kerangka Kerja Social Behavior and Change Communication (SBCC) Pada Komunikasi Kesehatan. Jurnal Pendidikan dan Sosial Humaniora. 2022;**2**(4):23-39

[27] Field EM, Maffioli EM. Are Behavioral Change Interventions Needed to Make Cash Transfer Programs Work for Children? Experimental Evidence from Myanmar. National Bureau of Economic Research; 2021

[28] Said-Mohamed R, Micklesfield LK, Pettifor JM, Norris SA. Has the prevalence of stunting in South African children changed in 40 years? A systematic review. BMC Public Health. 2015;**15**(1):1-10

[29] Kairiza T, Kembo G, Pallegedara A, Macheka L. The impact of food fortification on stunting in Zimbabwe:

Does gender of the household head matter? Nutrition Journal. 2020;**19**(1):1-12

[30] Chui KT, Alhalabi W, Pang SSH, de Pablos PO, Liu RW, Zhao M. Disease diagnosis in smart healthcare: Innovation, technologies and applications. Sustainability. 2017;**9**(12):2309

[31] Yuliansyah H, Sulistyawati S, Mulasari SA. Technological innovation is needed to accelerate stunting reduction in Indonesia. Epidemiology and Society Health Review (ESHR). 2022;**4**(2):89-90

[32] Manggarani CA, Ahmad ZA. The determining factors of student intention to donate in social crowdfunding media. Hasanuddin Economics and Business Review. 2020;**4**(1):7-14

[33] Abduh M. The role of Islamic social finance in achieving SDG number 2: End hunger, achieve food security and improved nutrition and promote sustainable agriculture. Al-Shajarah: Journal of the International Institute of Islamic Thought and Civilization (ISTAC). 2019;**2019**

[34] Palutturi S, Syam A, Asnawi A. Stunting in a political context: A systematic review. Enfermería Clínica. 2020;**30**:95-98

[35] Indonesia U. Strategi Komunikasi Perubahan Sosial dan Perilaku: Meningkatkan Gizi Remaja di Indonesia. Jakarta: UNICEF Indonesia; 2021

[36] Dearden K, Mulokozi G, Linehan M, Cherian D, Torres S, West J, et al. The impact of a large-scale social and behavior change communication intervention in the Lake zone region of Tanzania on knowledge, attitudes, and practices related to stunting prevention. International Journal of Environmental Research and Public Health. 2023;**20**(2):1214

Chapter 8

Animal Models in Myometrial Activity Research: Morphofunctional Features, Role of Oxytocin

Naira G. Hunanyan, Knarik V. Kazaryan and Tatevik A. Piliposyan

Abstract

One of the main functions of the reproductive system is providing the physiological process, which occurs by caudal spread of excitability of smooth muscle tissue and ensures delivery of the fetus. The results of this work confirm the importance of blood supply in automatism of the ovarian horn areas, which are the leading regions in propagation of electrical waves and coordination of other rhythmogenic loci. Morphofunctional methods have shown that the ovarian horn areas have strong vascularization, which is confirming the pronounced electrical activity of these loci. Oxytocin has a central role in changing of excitation patterns. Increasing hormone concentrations (10^{-2} μg/kg, 10^{-1} μg/kg, 1 μg/kg, 10 μg/kg) resulted in increase of the bursting activity duration of all studied myometrial areas. At the same time, rise in the frequency of spike rhythmogenesis was observed only at a dose of 1 μg/kg. Morpho-histochemical analysis revealed the existence of atypical cells with a high level of Ca^{2+}-dependent acid phosphatase in both distal rhythmogenic ends of the horn. However, the ovarian horn area had the greatest enzymatic activity. Thus, the obtained data give good reason to conclude that the ovarian horn area has a leading role in the myometrium.

Keywords: uterus, ovarian horn area, cervical horn area, spontaneous activity, oxytocin action, uterine artery ischemia, pacemaker activity, myometrium

1. Introduction

In all varieties of human internal organs, the uterus has a unique position not only because of differences in the structure and response to various environmental factors but also because of its special function compared to other visceral organs. The main functional importance for uterus has the coordinated contractile activity creating conditions for an orientation of waves to the uterine cervix [1, 2]. It has also been known that it is the spontaneous electrical activity is associated with uterine contractility.

Premature and also pathologically proceeding childbirth can lead to serious consequences up to the death of the fetus and mother. One of the possible methods of their prevention is the ability to control initiated uterine contractility. In this regard, the study of mechanisms providing this process will help to find out ways the solution this problem. Especially, it is important to reveal both the rhythmogenic regions and the driver pacemaker area providing the coordinated contraction of the organ. Several authors [3, 4] have noninvasively recorded the human uterine electrical activity from the abdominal surface, which allows the monitoring of parameters and their changes during rhythmogenesis, including discoordination of automatism, which leads to pathological consequences. Experimental resulting obtained in animal models, particularly, in rats, can provide a basis for similar clinical development.

The spontaneous electrical activity of nonpregnant rats is registered both from the uterine corpus and different areas of uterine horns and consists of intermittent bursts of action potentials resulting from the cyclic depolarization of the cell membranes [2, 5–7]. During pregnancy, especially in the later period, significant changes in generation of spontaneous activity in the rhythmogenic regions are observed, stimulating and coordinating the contractile activity of the uterus [3, 5, 8]. The study of the electrophysiological properties of all types of pacemaker activity presented in this organ will help to reveal the mechanisms providing their coordination during labor.

It is known that for normal functioning of the tissue optimal blood supply is necessary. Particularly, certain dependences of the electrical activity and, consequently, the process of delivery from a bloodstream are revealed for uterine smooth muscle tissue. Moreover, during pregnancy, occlusion of blood vessels [9, 10] is observed that also can matter in clinical practice. Based on this, morphofunctional research on studying the blood supply process in rhythmogenic regions of the uterus, providing generation of its contraction is necessary, which has not been presented in the literature yet.

Pregnancy is preceded by hormonal changes in the body, giving rise to certain changes in the excitability of myometrium [11–14] and, consequently, in the mechanisms of genesis of spontaneous activity. The solution of the aforementioned problem is most convenient to carry out in the conditions providing their activation. Oxytocin belongs to the strongest stimulators of uterine contractility during the birth process. It leads to the depolarization of membrane in myometrial cells and increase of frequency of spike discharges [3]. Oxytocin is produced in the hypothalamus and released into the maternal blood. It is also produced in the uterine placenta in the later period of pregnancy, and its concentration increases before delivery [15, 16]. Despite such important role of oxytocin during pregnancy and childbirth, the mechanism of its influence on the pacemaker activity has not been fully investigated.

The purpose of this work was to study the morphofunctional characteristics of rhythmogenic regions in the rat myometrium, as well as the identification of driver pacemaker areas under the influence of oxytocin.

2. Materials and methods

It is well-known that the sensitivity of the myometrium to hormones and mediators is raised in pregnancy, which is affecting autonomous spontaneous rhythmogenesis [17]. This is the reason why the experiments were done on nonpregnant rats.

Experiments were carried out on female animals (200–250 g) narcotized intraperitoneally with Nembutal (50–55 mg/kg) under the conditions *in situ*. The experiments were acute, and at the end of the recording the animals were killed. The peritoneal cavity was opened, and the uterine corpus with the uterine horns from two sides was exposed. The uterus was denervated by sectioning the nerves (plexus hypogastricus, uterinus, uterovaginalis). Spontaneous electrical activities were recorded simultaneously from the uterine corpus, uterine cervix, ovarian, and cervical ends of uterine horns. The activity of the uterine cervix was recorded by inserting a monopolar silver electrode into this area. Spontaneous electrical activity of the remaining areas was recorded from the surface of these regions with bipolar electrodes (interelectrode distance was 2 mm). The registration areas are schematically shown in **Figure 1**.

Oxytocin (5 units/ml, Biolek, Ukraine) was diluted in distilled water and injected intravenously at various doses: 10^{-2}, 10^{-1}, 1, and 10 μg/kg. Depending on the animal weight, such concentrations were possible to administrate by the introduction of different injection volumes — from 0.3 to 0.2 ml. Only one dose of oxytocin was studied on each animal in particular experiments.

Ischemia of the uterine artery was performed by an elastic cord prior to the ovary and its corresponding horn.

The results were recorded without a Faraday cage, and the level of valid signal was 10 μV. Statistical analysis of the data was performed by LabView, Origin8.5, and DIAdem programs.

Corresponding norms of parameters were registered and analyzed in each subsection of the present work, considering seasonal changes in the activity characteristics.

In order to study the morphofunctional properties of cellular structures the activity of Ca^{2+}-dependent acid phosphatase has been revealed [18]. This methodological approach is based on the detection of intracellular phosphorous-containing

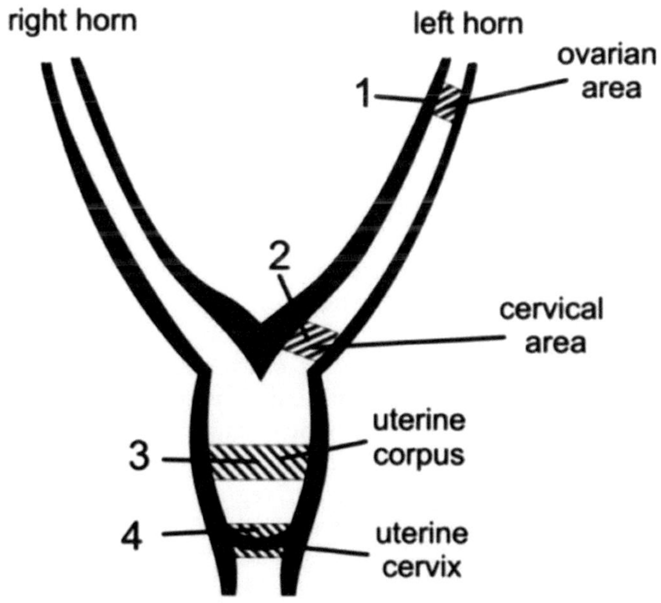

Figure 1.
Scheme of electrical activity registration in the ovarian horn area (1), cervical horn area (2), uterine corpus (3), and uterine cervix (4).

compounds taking key positions in the energy exchange processes directed on the preservation and self-reproduction of vital systems.

The vascular system of the rat uterus was detected using the calcium adenosine triphosphate histoangiological method of Chilingaryan [19]. The method is based on the selective phosphorus deposition segregated from ATP by calcium ions. Subsequently, the reaction product is converted to black lead sulfide. This method provides a clear selective detection of vascular-capillary network, and the major advantage is the possibility of simultaneous differentiation of the various links of the microcirculatory bed (arterioles, capillaries, and venules).

3. Results and discussion

3.1 Electrophysiological characteristics of different areas of the uterus and uterine horns

It is known that the electrical activity can be registered from both pregnant and nonpregnant uterus, but these activity discharges are scarce and have shorter duration in nonpregnant organisms [2, 3, 20].

According to the detailed electrophysiological analysis, rat uterine horns also have an ability to generate spontaneous electrical activity. Pacemaker areas are located in the ovarian and cervical ends of the uterine horn [21]. The uterine cervix also has autonomic spontaneous rhythmogenesis, which is completely asynchronous with activity of the uterine corpus in nonpregnant rats and is presented by slow-wave oscillations in membrane potential [2, 22, 23].

The uterus is unique among smooth muscular organs in that, it has an exceptional selectivity not only to various types of mediators and ions but also to sexual hormones. As noted above, the sensitivity of this tissue to hormones considerably increases during pregnancy, and especially in its late stages [11–13], which, naturally entails changes in the generation of electrical activity and, consequently, in its parameters. In these conditions, coordination of all presented types of autonomous spontaneous electrical activities is observed, which leads to the realization of peristaltic activity in the uterus [2, 8]. To reveal the synchronization mechanisms of rhythmogenesis in different areas of this organ, a prior analysis of the electrophysiological characteristics of these activities is necessary. The study of abovementioned issues would be best to carry out on nonpregnant organisms; in this case, the activity of the uterus and uterine horns can be taken as the norm.

In the present work, with simultaneous registration of activity in different areas of the organ, we tried to analyze the characteristics of the spontaneous electrical activities of nonpregnant uterine corpus, terminal ends of uterine horns, and uterine cervix.

The study of electrical activities of both the uterine corpus and different areas of the left uterine horn allowed us to register spontaneous electrical burst discharges (**Figure 2 (1-3)**). Electrical activity in aforesaid regions is presented by typical spike activity bursts that occur against the rather unstable level of membrane potential and last 41.8 ± 3.3 sec (n = 10), 45 ± 2.1 sec (n = 7), and 34 ± 1.5 sec (n = 10) respectively, in the ovarian and cervical ends of horn, uterine corpus, and then disappear.

In contrast to the described areas, the activity of the uterine cervix was usually recorded during all times of registration and presented by slow-wave rhythmic oscillations in membrane potential (**Figure 2 (4)**). The amplitude and frequency of activity corresponded to 189.1 ± 17.8 μV and 17.4 ± 0.7 oscill./min (n = 10) in this

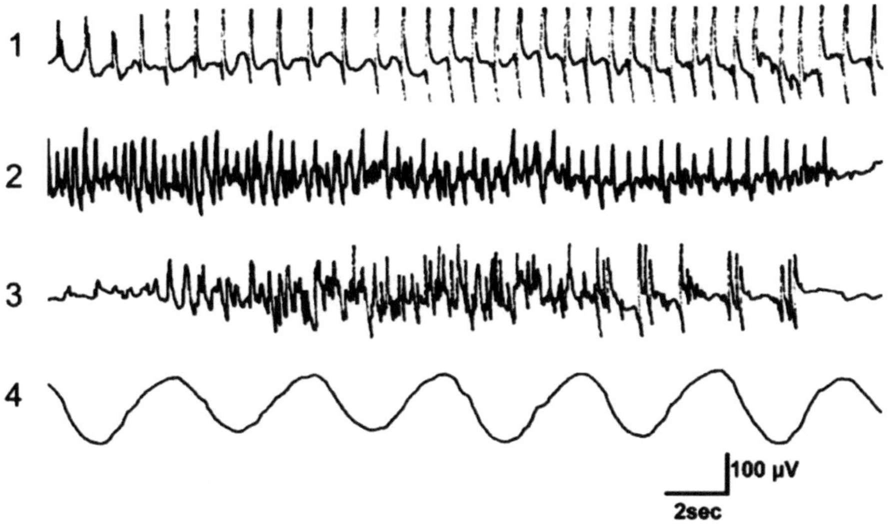

Figure 2.
Simultaneous registration of spontaneous activity in the studied uterine areas. The numbers on the left correspond to the regions of electrical activity registration presented in ***Figure 1*** *(n = 9).*

region. Based on our results and literature data, we can conclude about the autonomy of pacemaker activity rhythmogenesis of the uterine cervix.

Spike activity discharges in the three upper regions do not occur simultaneously in most experiments. But they work together in a certain period of time. Based on the registrations carried out continuously for 30 minutes, we have calculated the summarized periods of the active states of the uterine three upper regions (**Figure 3**). Despite the spike activity discharges lasted various time intervals, their total duration was almost the same for the cervical end of the horn and uterine corpus (10.95 ± 1.54 min and 11.8 ± 1.24 min, respectively (**Figure 3 (2,3)**). However, this parameter of activity for the ovarian locus of uterine horn was much greater than the above values (19 ± 1.12 min) (**Figure 3 (1)**).

There was a slow-wave rhythmic oscillation of the membrane potential in the uterine cervix, which has been observed during the entire period of registration.

According to the presented three upper curves in **Figure 2**, pacemaker activities of the uterine corpus and two regions of the uterine horn have the characteristics of a typical spike rhythmogenesis. Since activities of these regions alternate one another, it is difficult to establish which of them is primary in relation to the other. At the same time, comparison of these activity characteristics would help to establish the identity of mechanisms of their genesis.

The next series of experiments were carried out to examine the relationship between the frequency and amplitude of spike activities in the 1, 2 and 3 areas, presented in **Figure 1**. This correlation has an exponential characteristic in the uterine corpus (**Figure 4**). Changes in the frequency of spike activity genesis of given area are in the range of 30–200 spikes/min. Interesting is the fact, that in the range of 45–60 spikes/min frequency changes the same value of amplitude was registered (about 188.2 µV). It should be noted that these parameters are in the range of 45–80 spikes/min frequency changes, the probability of occurrence of which could be considered rather high for the uterine corpus (49% of cases), since the range of frequency changes is large.

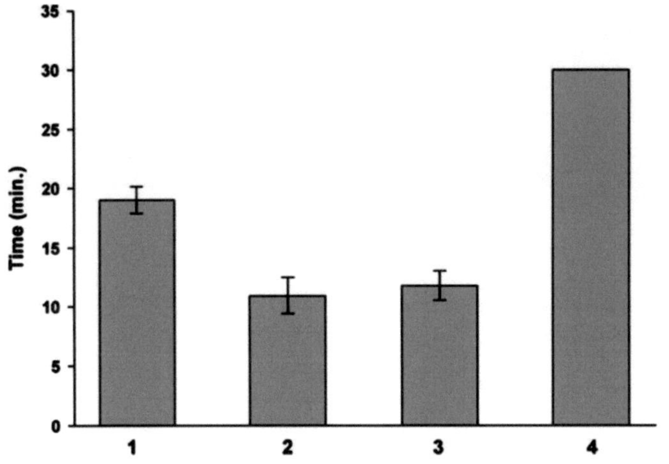

Figure 3.
Duration of the active state registration in the various areas of uterine horns and uterus. The numbers below correspond to the regions of electrical activity registration, presented in ***Figure 1*** *(n = 9).*

The amplitudes of spike activities are linearly dependent on the rhythm of their genesis in the cervical end of uterine horn (**Figure 5**). Changes in the frequency of automatism were almost in the same ranges (30–210 spikes/min), which corresponded to the uterine corpus activity. However, the abovementioned frequency range corresponding to the same value of the spike amplitude (≈188.2 μV) is a little shifted (55–80 spikes/min) and again in the aforementioned range of the most common frequencies (60.7%). As can be seen from comparison of the graphs presented in **Figures 4** and 5, the spike amplitudes decreased 2.4 times in both areas in the range of all frequencies. Despite a number of identical parameters of presented graphs, the relationships in these areas have different characters.

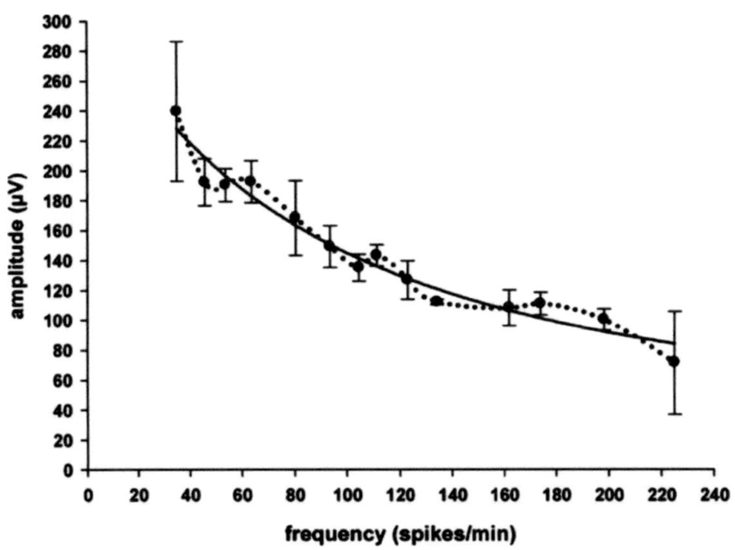

Figure 4.
The spike activity amplitude changes in the uterine corpus, depending on the frequency of its genesis. The dotted line connects all the mean values of the experimental data (n = 9).

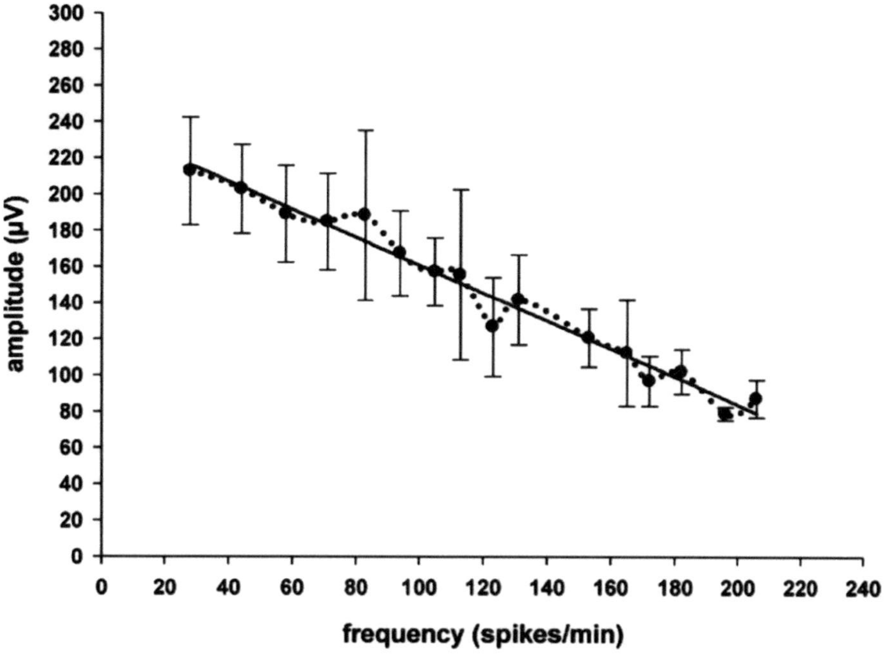

Figure 5.
The spike activity amplitude changes in the cervical horn area, depending on the frequency of its genesis. The dotted line connects all the mean values of the experimental data (n = 9).

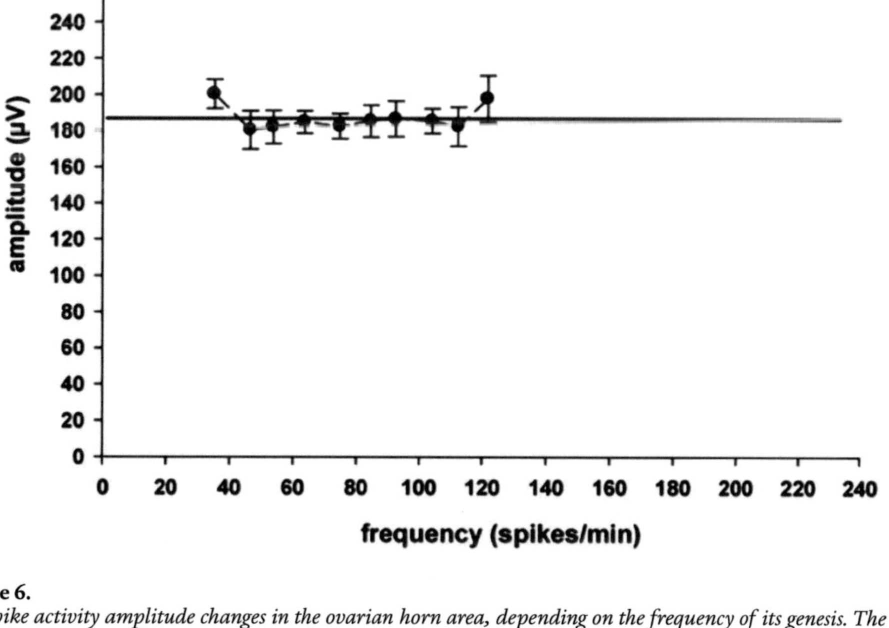

Figure 6.
The spike activity amplitude changes in the ovarian horn area, depending on the frequency of its genesis. The dotted line connects all the mean values of the experimental data (n = 10).

The correlation between amplitude and frequency is rather different in the ovarian part of horn compared to the uterine corpus and cervical part of uterine horn (**Figure 6**). Here the frequency of spikes is less scattered and mainly involves the range of 30–120 spikes/min. At the same time, the spikes having stable amplitude (186.8 ± 2.13 μV) are registered in the indicated range of frequency changes. In contrast to this area, similar amplitude value of spikes is observed within the most met frequency range (45–80 spikes/min) in the cervical end of horn and uterine corpus. Noteworthy is the fact that the frequency parameters of the same range of changes are also the most met (48.5%) in the ovarian end of horn such as the cervical end of horn and uterine corpus.

Thus, despite the obvious differences in the amplitude and frequency correlations between ovarian end of horn and downstream areas, which are also generating spike activity, the greatest probability of occurrence of frequencies corresponds to the same range of changes for all three areas.

3.2 Blood supply as a factor regulating pacemaker activity of the myometrium

It is known that the spontaneous electrical activity of smooth muscle tissue is myogenic in nature [2, 6, 21, 24]. If the myogenic rhythm plays the main role in occurrence of the peristalsis in smooth muscle, the humoral control is necessary for modulation and coordination of the contractility patterns [25]. Particularly, the blood supply due to its transport function plays an important role in this process. Thus, smooth muscle of the gastrointestinal tract is rather resistant to decrease in blood flow [26]. At the same time, anoxia rapidly decreases the rat uterine contractile activity and its complete suppression takes place within a few minutes under conditions of ischemia [27–29].

It is also known that parameters of the smooth muscle activity can vary within a rather wide range not only suggesting about the relationship of automatism and blood flow but also confirming the variability of this process [30].

Rats and mice have a duplex uterus (two separate uterine horns). The loops of the uterine artery encircle each horn separately [27, 31], while the blood flow in the right and the left horn is realized in two directions: from ovarian and cervical ends to the center [32, 33].

According to a detailed electrophysiological analysis, unlike the uterus, any part of which is able to generate spontaneous discharges, pacemaker areas are located in terminal ends of uterine horns in rats [21, 34]. The results of the previous section confirm the noticeable differences of characteristics of spontaneous activity in the ovarian end of uterine horn and downstream rhythmogenic areas. Moreover, it is shown that the total duration of generation of the electrical bursts is much higher in this area. However, there are very few special studies in the literature devoted to the experimental analysis of blood supply in the ovarian region and its importance in electrical activity generation. The following paragraph analyzes these issues. The uterine horns and the uterine artery loops that feed each horn are presented schematically in **Figure 7**. The areas of activity registration are also shown (respectively numbered). In the first series of experiments, a comparative analysis of changes of spike activity parameters in the uterine corpus and terminal parts of uterine horns was carried out under conditions of the uterine artery ischemia. **Table 1** presents the values of the parameters of rhythmogenesis of the noted areas in norm.

About 20-min ischemia of the uterine artery, feeding the ovarian part of horn, led to certain changes in the activity parameters of each area, respectively (**Figure 8**). For illustrativity, all results are presented in percents related to norm.

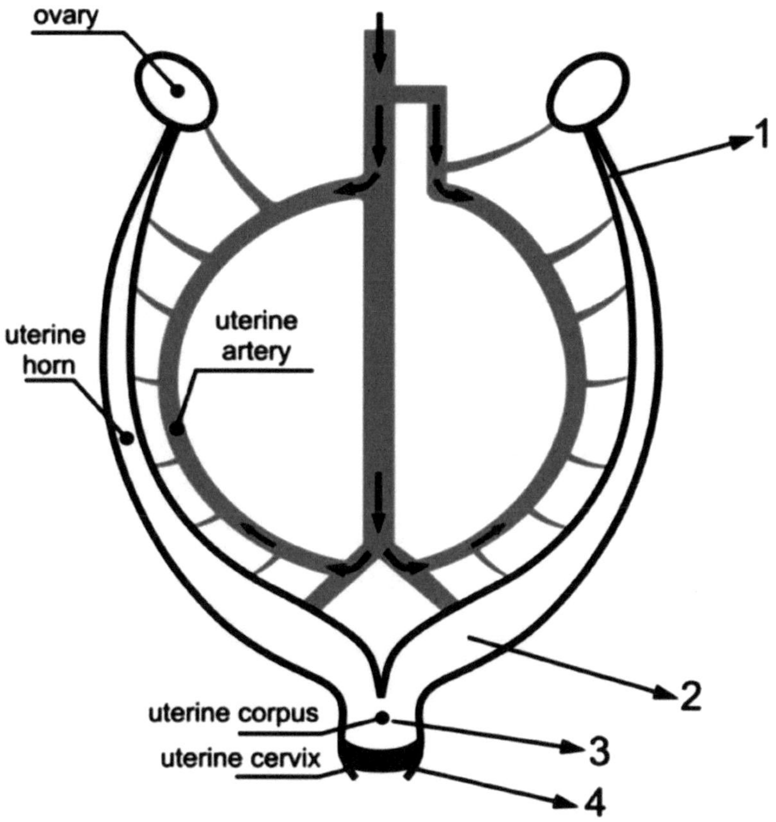

Figure 7.
The blood supply scheme of the uterus and uterine horns in rats. The numbers on the right correspond to the regions of activity registration: 1-ovarian end of horn, 2-cervical end of horn 3-uterine corpus, and 4-uterine cervix.

Areas of registration	Amplitude of spikes, μV	Frequency of spikes, spikes/min	Duration of activity discharges, sec
1	142.3 ± 4.99	84.5 ± 6.16	36.3 ± 2.66
2	126.0 ± 2.00	92.5 ± 2.29	34.9 ± 1.79
3	104.7 ± 2.40	116.9 ± 4.83	19.6 ± 4.94

Table 1.
*Parameters of spike activity in the uterine corpus and terminal parts of uterine horn in norm. 1, 2, and3 are the corresponding registration areas (**Figures 1** and 7).*

As can be seen in **Figure 8(A)**, the amplitude of spikes in the ovarian horn area undergoes to the greatest changes. In the cervical end of horn and uterine corpus, this parameter decreases by less than a quarter of its initial value (24.4 and 23%, respectively). Interestingly, suppression levels of the amplitude in these areas are similar. At the same time, a different picture is observed for the ovarian area of uterine horn. There is a sharp, almost twice decrease in amplitude (to 54.5%).

According to **Figure 8(B)**, the frequency parameters of all three areas compared to the amplitude parameters are less subjected to changes under conditions of the ischemia. In the cervical horn area and uterine corpus, this parameter changes

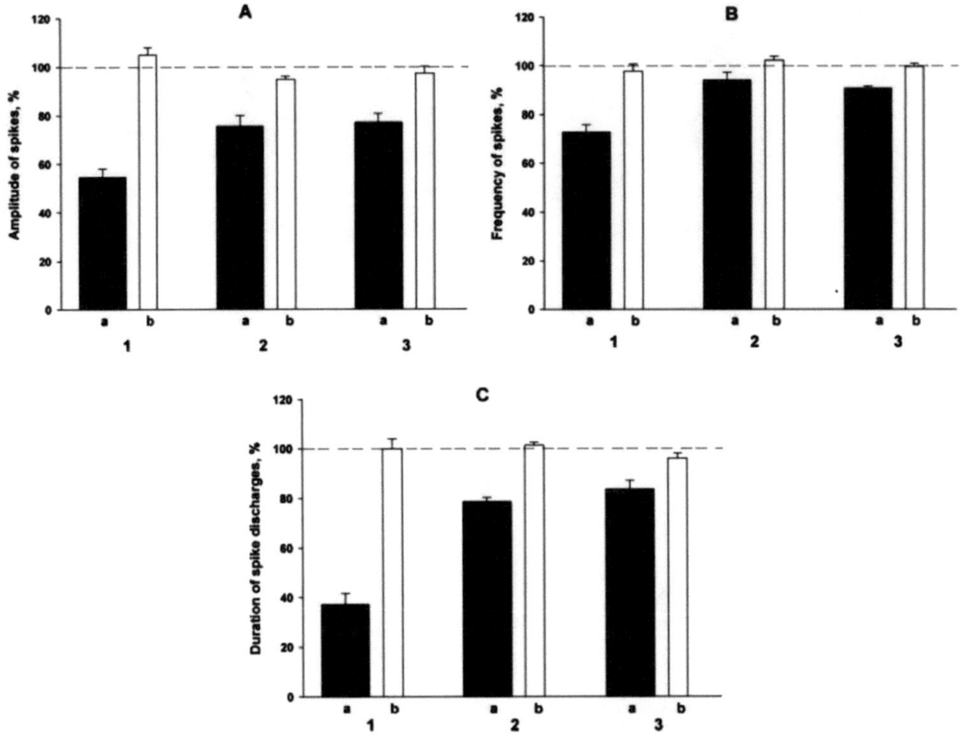

Figure 8.
The influence of ischemia on the parameters of spike activity in the uterine corpus and terminal areas of uterine horn. A - amplitude of the spikes, B - frequency of the spikes, C - duration of the spike discharges. a - uterine artery ischemia; b - restoring blood flow. The numbers below correspond to the regions of electrical activity registration presented in ***Figures 1*** *and 7 (n = 8). The dotted line demonstrates the norm.*

insignificantly (by 5.9 and 9.3%, respectively), but decrease in spike frequency reaches to 27.2% in the ovarian horn area.

A similar trend of changes in activity parameters is also shown for the duration of the spike discharges in the studied areas (**Figure 8 (C)**), in the proximal horn area, the duration of burst genesis was significantly shortened (by 62.8%), while decrease in this parameter for two subsequent areas (**Figures 1** and 7 **(2,3)**) corresponded to only 21.2 and 18.3%. Restoration of blood flow after ischemia, as a rule, increased all parameters to near-normal values.

Despite the significant suppression of activity in the ovarian area, its full inhibition was not observed. Probably, the total block of rhythmogenesis could be achieved by prolonging time of ischemia or increasing force of the cord pressure to the complete closure of the vessel lumen. But such experiments could damage the artery.

As noted above, blood enters the uterine horn from ovarian and cervical ends and flows to the middle region of it. Blood supply of the uterine corpus and uterine cervix is provided by an additional branch of the artery (**Figure** 7) [27]. Ischemia of the uterine artery, which enters the ovarian horn area, entails significant changes mainly in the activity characteristics of this region. A little inhibition of parameters in the cervical end of horn and uterine corpus under conditions of such ischemia may indicate that there is a likely certain connection between the studied rhythmogenic loci. Actually, if the cervical horn area had an autonomous blood supply, the activity parameters would remain unchanged.

Thus, under the above experimental conditions, the main rhythmogenic region of uterine horn undergoes the greatest activity inhibition compared with the below located pacemaker areas.

In the next series of experiments, we analyzed the activity parameter changes in the uterine cervix under similar conditions of ischemia (**Figure 9**). As mentioned above, activity of the uterine cervix is completely different from that of all other areas of this organ. In norm, slow-wave sinusoidal-type activity is registered in this area with a frequency of 15.1 ± 0.34 oscill./min and amplitude of 282.1 ± 14.5 μV, which is fully autonomous. Even some authors have described the cervix as an entirely separate from uterus organ [2, 35, 36].

As can be seen by comparing **Figures 8(A)** and **9(A)**, the opposite changes in the amplitudes of the ovarian horn area and uterine cervix were noted under conditions of ischemia. Twice decrease in the spike amplitudes of the horn is accompanied by one and half time increase in the slow-wave amplitudes of the uterine cervix (up to 152.3%). The frequency characteristics of activities in the abovementioned areas in both cases decreased and cervical rhythmogenesis slowed by only 14.4% compared to the norm (**Figure 9 (B)**). All changes in characteristics of the slow-wave activity occur against a background of its continuing genesis.

It is known that slow-wave oscillations of the membrane potential recorded from the uterine cervix in norm provide its closure by the involvement of circular muscle layers [2]. Significant increase in amplitude of the cervical waves in ischemia indicates the activation of this process that stimulates the cervical closure. During the labor, the longitudinal muscle layers of the cervix, which are able to generate spike activity, are involved in the longitudinal polar contractions provided by the main rhythmogenic region (ovarian area of the horn).

Thus, there is a certain connection between rhythmogenesis in the ovarian horn area and uterine cervix. Probably, the results obtained in this work related to the opposite changes in the activity parameters of the terminal uterine areas under

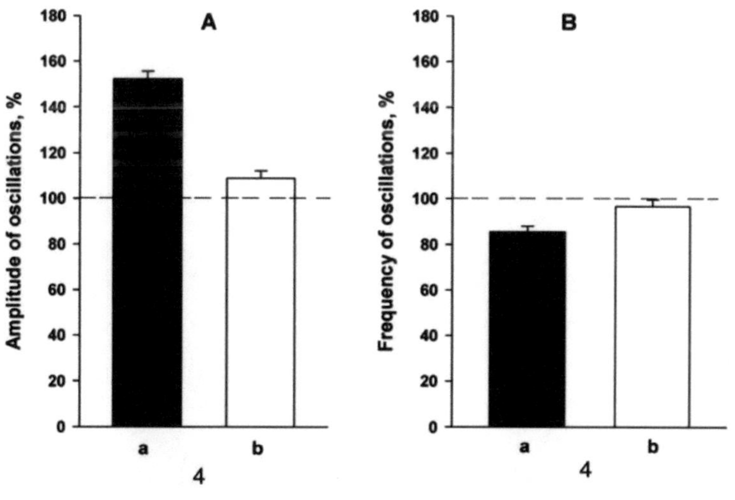

Figure 9.
The influence of ischemia on the parameters of slow-wave activity in the uterine cervix. A - amplitude of the waves, B - frequency of the waves. a - uterine artery ischemia; b-restoring blood flow. The numbers below correspond to the regions of electrical activity registration presented in ***Figures 1*** *and 7 (n = 8). The dotted line demonstrates the norm.*

conditions of the ischemia can be explained by the presence of such interaction between them [37].

In the next series of experiments, the microcirculatory bed of the ovarian region was investigated. In parallel with the vascularization of aforesaid area, the vascular bed of the ovarian nearby area and of distally located middle horn area were also studied. A strong vascularization with highly branched arterial system was detected on the longitudinal sections of the ovarian locus (**Figure 10(A)**). Therefore, there is a huge amount of the microvessel cross sections in the slices. The transition zone between the ovaries and uterine horn is characterized by less twisted blood vessels (**Figure 10(B, C)**). In comparison with the mucous and serous membranes, the muscular layer is highly vascularized in the horn area closest to the ovary. Blood vessels from the surface enter the uterine horn directly, almost without twisting, are assembled in a bunch (shown the blacked-out surface), and by this, way creates a strong vascularization of this region (ovarian end of the horn). These vessels take again a strongly twisting direction in the middle region of uterine horn. That is why there is a large number of their cross sections along with the short microvessels in the slices (**Figure 10(D, E)**).

The strong vascularization of the ovarian end of uterine horn is one of the possible factors, providing strongly pronounced automatism in this region. Vessels take a twisting direction in the middle area of uterine horn, and therefore their close contact with the tissue is less expressed. Such blood vessel location

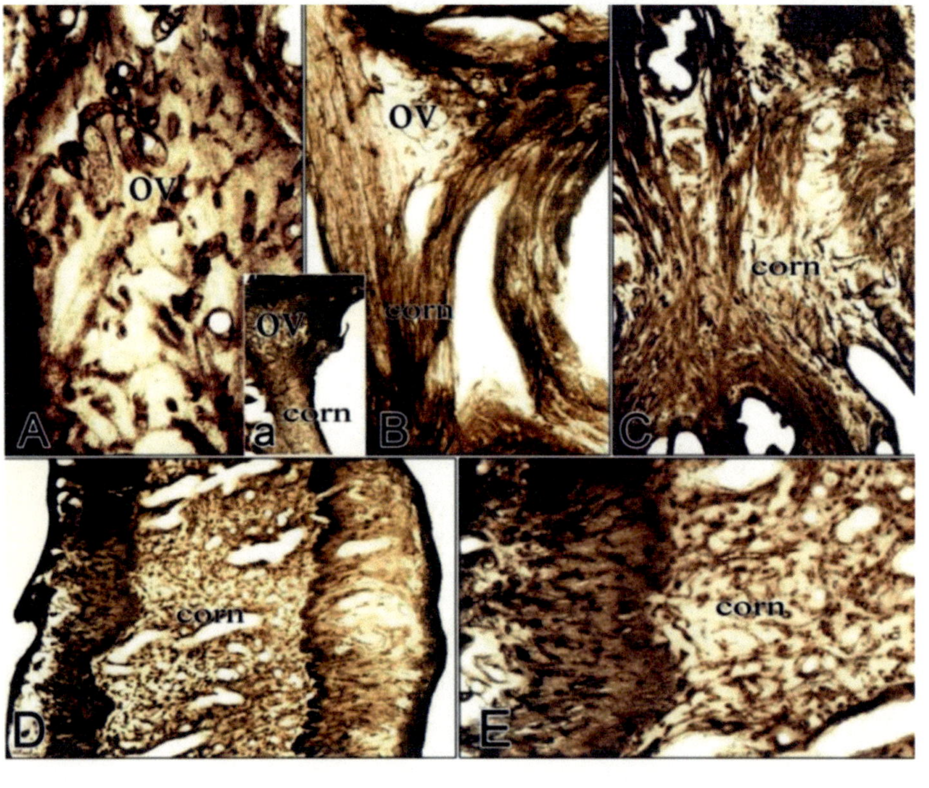

Figure 10.
Blood vessels of the microvasculature in the longitudinal sections of ovary (A, a), the initial region of the uterine horn (B, C) and the middle region of the uterine horn (D, E). ov-ovary; corn-uterine horn. Zoom: ocular 10, objective 2.5 (a), 6.3 (B), 10 (C, D, E), 16 (A).

throughout the uterine horn is specific to immature and nonpregnant females. Throughout pregnancy further development of the bloodstream and, respectively, the vascularization is observed. Branching in the vascular system occurs, providing blood flow to individual uterine segments associated with fetuses. Each uterine segment has an abundance of closely located blood vessels (unpublished data). It can be assumed that during reproductive organ formation, there are appropriate stages of the circulatory system development. Perhaps such vascular system was developed in the process of evolution to provide individual feeding of each segment by the local automatism, though the latter is coordinated with the main rhythmogenic region.

3.3 Role of oxytocin in activation of spontaneous electrical activities of the uterine corpus and uterine horns

Oxytocin is considered as one of the main regulatory substances of contractile activity in the process of childbirth [38]. Early studies have shown that under the effect of oxytocin, membrane depolarization is observed in myometrial tissue, leading to membrane excitability and conductivity, which results in increased activity and the occurrence of contractile events [3, 39, 40]. For this reason, the given substance is easy to use for the identification and study of changes in the characteristics of spontaneous electrical activity, leading to its coordinated activity for the generation of the subsequent contraction.

It is known that smooth muscle tissue of myometrium, as well as the gastrointestinal tract and urinary tract, is characterized by ability to generate autonomous spontaneous activity [2]. However, there is very little data in the literature about the mechanisms underlying the generation of pacemaker activity in the uterus in contrast to above mentioned other types of smooth muscle formations in which the specialized interstitial cells of Cajal (ICC) have been found, providing generation and coordination of muscular activity [41, 42]. Some cells that are completely different from myocytes by their morphology and ultrastructure were also found in studies on uterine strips of pregnant rats [43]. The morphological studies concerning the identification of atypical cells in myometrial tissue are easier to carry out under conditions providing the activation of spontaneous rhythmogenesis.

Research on the influence of oxytocin on the automatism of different areas of the uterus and uterine horns by electrophysiological and morpho-histochemical methods can help to tackle the abovementioned problems.

As already mentioned, unlike the uterine corpus, only the terminal parts (ovarian and cervical) of uterine horns can generate spontaneous electrical activity [44].

In the first series of experiments, we analyzed the fast spike processes observed in three upper regions, which are presented in **Figure 1**. We analyzed the duration of genesis of electrical discharges and spike frequencies within the bursts for the terminal parts of horns and uterine corpus (**Figure 1(1-3)**). These parameters can be ascribed to the number of the main characteristics determining the contractile activity of the organ [45].

It is known that non-regular bursts of spontaneous electrical activity can also be observed in nonpregnant rats [11, 22, 46, 47], and automatically arising activity discharges recorded in different areas of the organ have different duration and spike frequencies. **Table 2** presents our obtained data from the upper areas of the rat uterus in norm (**Figure 1(1-3)**). Interestingly, the duration of the activity recording and frequency of the spike genesis in the cervical ends of both horns a little exceed the same

Areas of registration	Duration of activity discharges, sec	Frequency of spikes, spikes/min
Ovarian end of the left horn	50.8 ± 2.74	73.7 ± 3.78
Cervical end of the left horn	55.2 ± 4.88	84 ± 4.11
Ovarian end of the right horn	28.3 ± 5.08	59.2 ± 3.41
Cervical end of the right horn	31.2 ± 3.61	75.7 ± 3.85
Uterine corpus	32.8 ± 2.00	92.1 ± 4.27

Table 2.
Parameters of spike activity in various areas of uterine horns and uterus in norm.

parameters in the ovarian ends. In addition, the right uterine horn is less active in norm as it is characterized by significantly lower values of these parameters compared with the left horn.

In the next series of experiments, we studied the durations of the discharge genesis and spike frequencies in the uterus and terminal parts of horns, depending on the concentration of oxytocin. For illustrativity, all results are presented in percents related to norm. The presence of oxytocin in any concentration causes a significant increase in the duration of genesis of electrical discharges in the both ends of uterine horns, as well as in the uterine corpus itself (**Figure 11**). In the ovarian ends of both the right and the left horns (**Figure 1(1)**), the maximal increase (almost threefold) of this parameter was noted at the oxytocin dose of 10^{-1} μg/kg. In the cervical ends of horns, closest to the uterus (**Figure 1(2)**), as well as in the uterine corpus itself (**Figure 1(3)**), the longest duration of electrical bursts is observed at the oxytocin concentration of 1 μg/kg. It is important to note that the subsequent increase in the oxytocin concentration to 10 μg/kg led to a shortening of duration of the active state registration. Nevertheless, this parameter significantly exceeds the values in norm for all the studied areas. Interestingly, the spontaneous bursts of activity, as a rule, were not recorded in the ovarian end of the right horn in these conditions.

Thus, by increasing the oxytocin concentration in the considered limits, we have shown a significant increase of the active state duration in all studied areas.

A somewhat different picture is observed in the study of frequency parameters of the spike rhythmogenesis. According to the data of **Figure 12**, for the uterine horns and uterine corpus, the greatest increase of the spike frequency in discharges is revealed at the oxytocin concentration of 1 μg/kg. In contrast to the duration of the activity discharges (**Figure 11**), the presence of oxytocin in a concentration of 10^{-2} μg/kg caused some decrease in the frequency of the spike rhythmogenesis in terminal parts of uterine horns. The increase in the oxytocin concentration to 10^{-1} μg/kg led to an acceleration of the spike rhythmogenesis. A significant increase in the spike frequency was registered in the ovarian end of the right uterine horn at the oxytocin dose of 1 μg/kg; at the same time, in the cervical areas of both horns the frequency parameters did not undergo significant changes with increase of the administrated oxytocin concentration (**Figure 12(a, b)**).

In the uterine corpus, a small increase in the spike frequency was observed within the limits of the oxytocin concentration changes from 10^{-2} to 1 μg/kg, while at 10 μg/kg, this parameter decreased to the level slightly lower than the norm (**Figure 12(c)**).

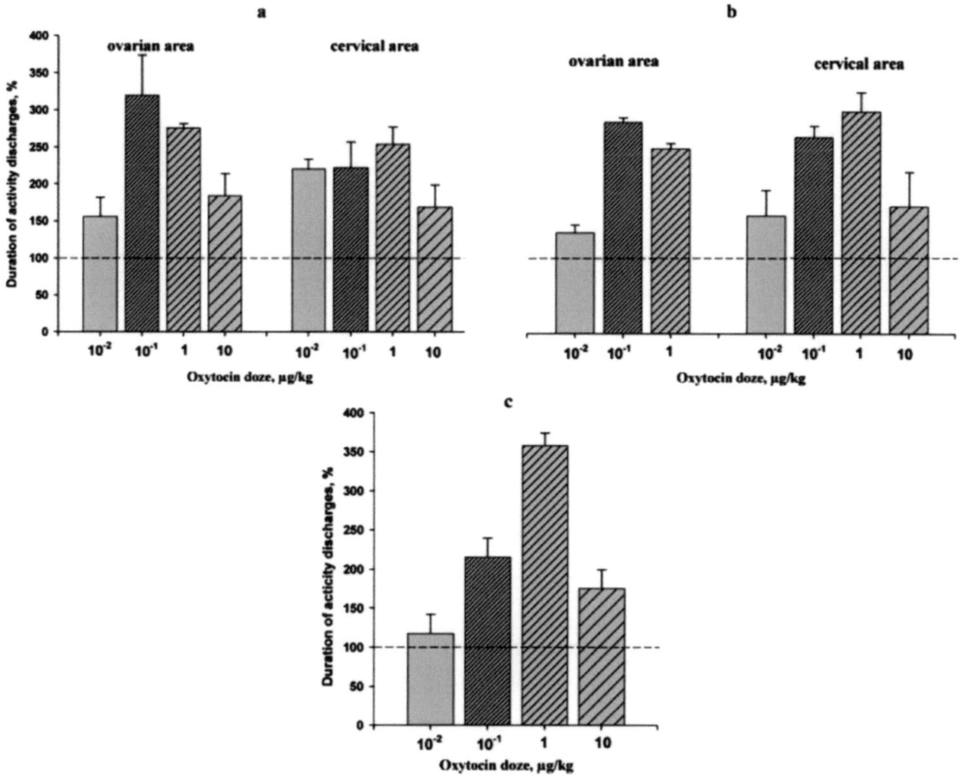

Figure 11.
The influence of oxytocin on duration of genesis of the spike activity discharges in the uterine horns and uterine corpus. The norm is shown by the dotted line (n = 19). a - the left horn, b - the right horn, and c - the uterine corpus. Oxytocin concentrations are marked by different shadings.

Small differences between the values of the spike frequency in the ovarian ends of the right and the left horns seem to be due to the high values of this parameter in the norm (**Table 2**). Unlike the duration of the genesis of the activity discharges, the frequency of the spike automatism in the studied areas is less affected by the action of oxytocin.

The uterine cervix under the influence of this agent behaved quite differently (**Table 3**). At the oxytocin concentration of 10^{-2} µg/kg, no changes were observed in the activity characteristics of this zone, only slow waves were recorded with the frequency corresponding to that in norm (16.6 ± 1.52 oscill./min). The increase in concentration of oxytocin to 10^{-1} µg/kg resulted in the disappearance of the waves, and spike activity was recorded during a certain period of time. The dependence of the duration of the spike generations and their frequency on the oxytocin concentration is presented in **Table 3**. The higher oxytocin concentration-10 µg/kg somewhat decreased these studied parameters of the uterine cervix such as the presented data for the remaining organ areas.

Comparing the effect of oxytocin on duration of the discharge genesis and frequency of the spike activity, it can be concluded that the cervical ends of uterine horns and the uterine corpus can be grouped together according to similar changes in these parameters with increase of the oxytocin concentration. The ovarian ends

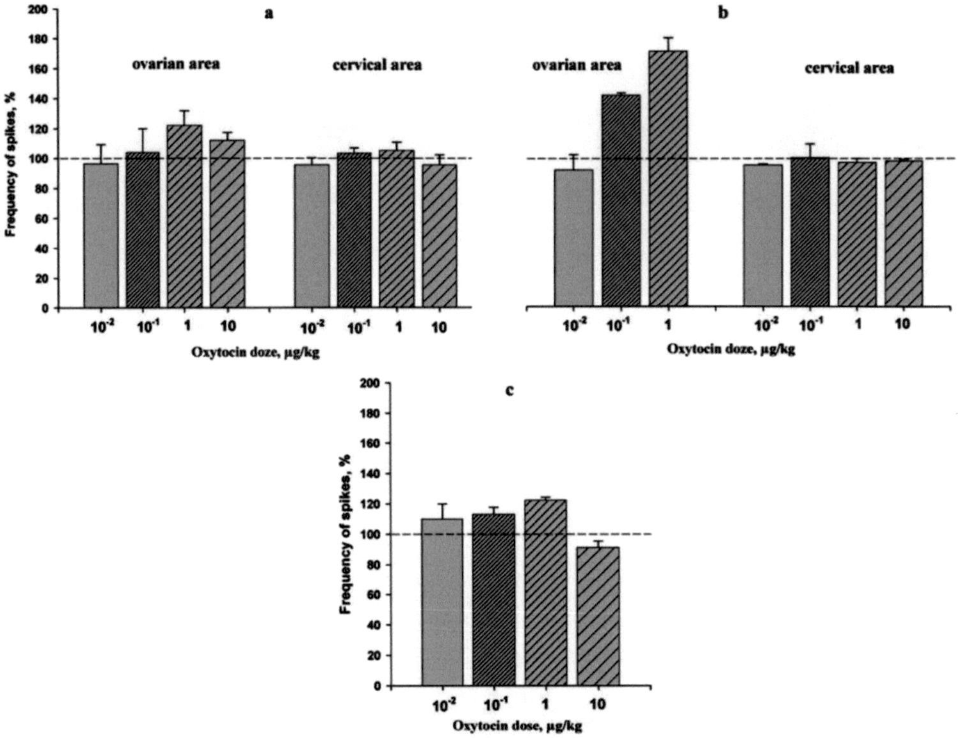

Figure 12.
The influence of oxytocin on frequency of the spike activity in the uterine horns and uterine corpus. The norm is shown by the dotted line (n = 19). a - the left horn, b – the right horn, and c – the uterine corpus. Oxytocin concentrations are marked by different shadings.

Concentration of oxytocin, µg/kg	Duration of activity discharges, sec	Frequency of spikes, spikes/min
10^{-2}	—	—
10^{-1}	50.0 ± 9.0	105.3 ± 18.7
1	66.5 ± 17.5	120 ± 12.0
10	25 ± 5.8	112.5 ± 16.8

Table 3.
Parameters of spike activity in uterine cervix under the effect of oxytocin.

of both uterine horns had somewhat different characteristics of changes in these parameters (**Figures 11** and **12**) within the same limits of the oxytocin concentration changes. Therefore, a certain interest is the comparative analysis of morphological pictures of all three upper regions presented in **Figure 1**.

Morpho-histochemical studies revealed that there are round or oval atypical cells with the high level of the Ca^{2+}-dependent acid phosphatase activity in longitudinal sections of the ovarian ends of uterine horns in intact rats (**Figure 13(Aa)**). The most clearly detected structures are cell nuclei. Gomori-positive round ovoid, coccobacillar, comma-like granulations are found on the light cytoplasmic background. Quite often these cells fuse with each other, therefore, many nuclei are revealed in the large mass of cytoplasm in the preparations. It is important to note that the greatest accumulation of these cells was detected in the ovarian ends of horns. The same cells were found in the cervical parts of horns (**Figure 13(Cc)**) and in the uterine corpus (**Figure 13(Ee)**). Unlike the atypical cells, the enzymatic activity of the acid phosphatase is significantly decreased in the typical myogenic elements of the uterine wall.

Under the oxytocin action, the enzymatic activity was sharply increased in the cytoplasm of atypical cells. In some foci, these cells were concentrated in clumps and their shape became variable, which amplified the staining intensity and giving to the location the blacked-out look. In the ovarian part of uterine horn, these cells were stained darkly, and it was impossible to determine the boundaries of the nucleus and the cytoplasm (**Figure 13(Bb)**). Atypical cells of the uterine corpus (**Figure 13(F)**) and of the cervical horn area (**Figure 13(D)**) were also intensively stained.

According to the literature data, the cells resembling the interstitial cells of Cajal were revealed in the upper urinary tract, particularly in the pyeloureteral anastomosis [43]. They were of irregular shape, with oval nucleus, and numerous contacting processes and were described as the fibroblast-like cells providing pacemaker activity.

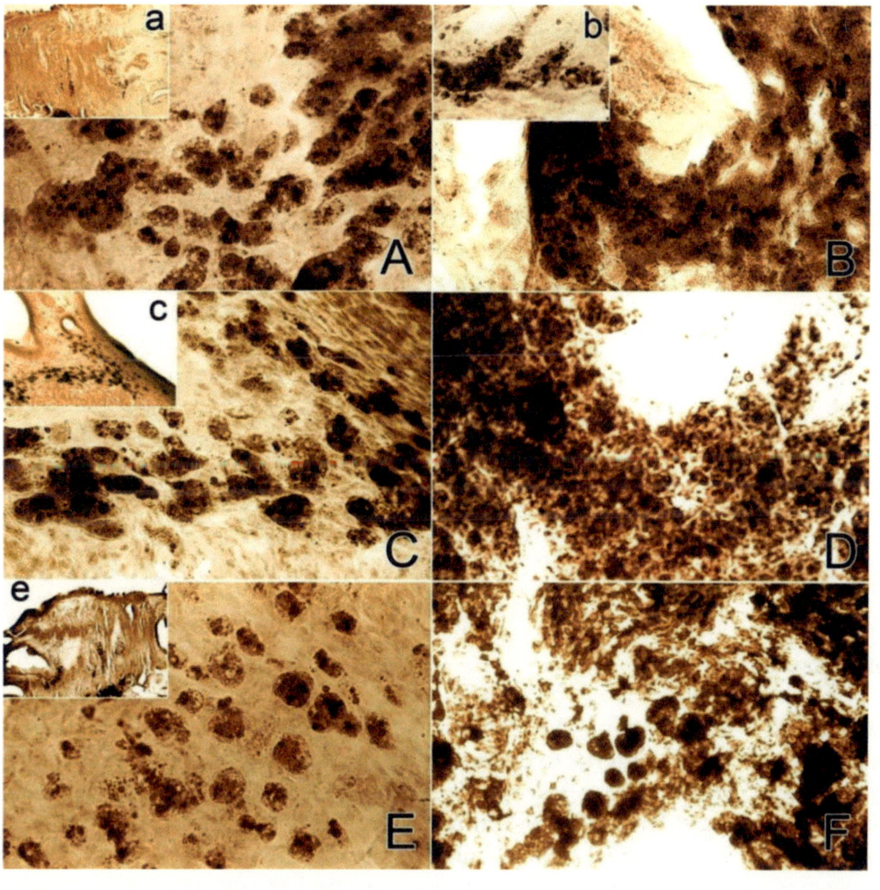

Figure 13.
Longitudinal sections of the uterine corpus and the various horn areas. Intact rats: A-ovarian area of the left horn; C-cervical area of the left horn; E-uterine corpus. At the oxytocin administration: B- ovarian area of the left horn; D-cervical area of the left horn; F-uterine corpus. Magnification: 25 × (a,c,e); 400 × (b); 1000× (A, B, C, D, E, F).

According to the obtained data, smooth muscle cells differed from the revealed atypical cells by the level of enzymatic activity of acid phosphatase. Thus, atypical cells had high Ca^{2+}-dependent acid phosphatase activity, although additional immunohistochemical markers are needed to determine the nature of these cells. Nevertheless, the ovarian part of horn is established to have the highest enzymatic activity of the acid phosphatase among all considered areas.

Our histochemical data confirm completely the presence of different "physiological" states in the studied organ areas. Thus, the complete correspondence of the electrophysiological and morphological results is observed.

It was shown that the duration of spike discharges is a determining factor in the generation of coordinated contractions in the uterus [48]. Analysis of the oxytocin action showed that the main effect was expressed as a significant increase of duration of the spike discharges in the studied areas of the myometrium. The rather low content of the substance in the blood (10^{-1} μg/kg) caused a significant increase in this parameter (up to threefold) compared with the norm. According to the literature data [45], oxytocin in a concentration of 2×10^{-2} μg/kg (≈0.2 nM) is sufficient to stimulate the spike activity, which is expressed by the increase of spike frequency, prolongation of discharge generation, and consequently, contraction periods. The further increase in the concentration of oxytocin causes gradual inhibition of these parameters [3, 45], which also completely agrees with our results obtained at the oxytocin concentration of 10 μg/kg not only for the uterine corpus but also for the ovarian and cervical ends of uterine horns.

Naturally, to provide the uterine contractile activity during the labor, coordination of all pacemaker areas, determining the polarity of the uterine direction, is necessary. The leading role in providing longitudinal contraction of the uterus is ascribed to the pacemakers located in the ovarian ends of uterine horns [34]. It is also shown that the duration of the electrical activity discharges is a determining parameter for the occurrence of coordinated contractions in the uterus [3]. According to the results presented in this work, it is the ovarian end of uterine horn, activation of which by oxytocin is found to be accompanied by the longest duration of genesis of the electrical discharges. Morpho-histochemical studies also confirm this fact — the greatest number of atypical cells has been found in this area, which is characterized by the highest acid phosphatase activity.

According to our preliminary results (data not shown), spontaneous electrical activity of the organ was not observed in 15–20-day-old female rats. Morphological studies also did not reveal above described atypical cells. We assume that certain rhythmogenic (atypical) cells develop during puberty, and they provide electrical activity of the uterus.

4. Conclusions

By definition Norwitz et al. [49], labor is the physiological process by which a fetus is expelled from the uterus to the outside world in the result of regular uterine contractions accompanied by cervical effacement and dilatation. It is known that uterine contractility is dependent on action potentials and their propagation along the tissue [50]. The frequency and duration of contractions are determined respectively by the frequency of the action potentials within a burst and the duration of a burst. The amplitude of contractions depends on the number of propagating action potentials [51].

The membrane potential value of myometrial smooth muscle cells is much lower than those of skeletal muscle and nerve fibers and is approaching to the threshold for spontaneous discharges. From this aspect, every cell in smooth muscle tissue can act as a pacemaker in certain conditions [52, 53]. Unlike the uterus, any part of which is able to generate spontaneous discharges [20], two rhythmogenic regions have been revealed in the uterine horns: the ovarian and cervical ends of the horns [21, 34]. The studies on nonpregnant uterus have shown that bursts of electrical discharges from the upper parts of this organ are propagated within a few mm [36]. Based on this, the spread of the excitation waves is observed in a very limited area from each rhythmogenic region and is absent along the horns. The study of the propagation rate of the electrical activity patterns along uterine horns in guinea pig showed a direct dependence on the electrical coupling between smooth muscle cells. At birth, the lowest resistance between them is observed [54].

According to recent studies, the uterine peristalsis could be provided by the association of the separate pacemaker areas in large rhythmogenic loci during labor [8]. Received data concerning such coordinated among themselves amplitude parameters of rhythmogenesis of the proximal part of horn do not exclude its leading role in providing the synchronization of spontaneous activity of the all organ. Based on the identity of detected ranges of the most met frequency parameters of action potentials with the same amplitude value (188 μB) in uterine corpus and both rhythmogenic regions of uterine horn, it is impossible to exclude the existence of coordination of their activities by the abovementioned manner.

The uterine cervix also has autonomic spontaneous rhythmogenesis. In the contrast to the uterine corpus and uterine horns, it is presented by slow-wave oscillations in membrane potential in nonpregnant individuals. The latter provides the closure of the cervical lumen due to circular muscle activity. In the late stages of pregnancy and during labor a totally synchronized spike activity of all pacemaker areas of the uterus is observed. In these conditions, the uterine cervix is able to pass spike electrical signals of proximal regions of the organ [2, 55].

Interestingly, according to the results presented in this work, ischemia of the uterine artery supplying blood to the ovarian end of uterine horn [56] leads to the opposite changes of amplitude parameters in the abovementioned region and uterine cervix. With twice decrease of amplitude value of spikes in uterine horn, this parameter of slow-wave activity in the uterine cervix increased one and a half times (to 152.3%). Thus, there is a certain correlation between rhythmogenesis in the ovarian end of horn and uterine cervix. As it was noted above, during labor occlusion of the uterine vessels is observed. Reduced blood supply (from 30 to 100%) leads to relaxation with intermittent increasing contractions during labor [21, 30]. Probably, significant increase of amplitude of slow waves in the uterine cervix under the influence of ischemia confirms the activation of that process, which stimulates compression of the cervical lumen during suppression of spike automatism in the proximal locus of horn, leading to the muscle relaxation. Possibly, this mechanism provides feedback for the normal process of delivery. The results presented here may be important for the development of a model, allowing to regulate the uterine contractility in pathological situations in clinics.

In analysis of the oxytocin action on the ovarian areas of uterine horns, it can be noted that the main effect under the influence of this hormone is expressed as a significant increase of duration of spike discharges. At the same time, rather low content of the substance in blood (10^{-1} μg/kg) caused significant (up to threefold) increase of this parameter compared to the norm. To provide the uterine contractile activity during the labor, coordination of all pacemaker areas, determining polarity of its direction, is necessary. According to literature data [34], as well as the results presented in this work, the coordinating role can be assigned to the ovarian end of horn. It is also shown that the duration of electrical activity discharges is a determining factor of the occurrence of the coordinated uterine contractions [3, 57]. The longest duration of the genesis of electrical discharges in the ovarian end of horn under the stimulation by oxytocin also confirms the leading role of pacemakers of this region. In this case, a simple registration of duration of the activity genesis from the abdominal surface of patients not requiring additional measurement of the amplitude and frequency values can help in clinics for an objective assessment of coming labor.

As it was mentioned above, nowadays the abilities of clinicians to regulate the uterine contraction, particularly, in preterm labor, remain limited. Possible solution of this problem may be the inhibition of contractions by regulation of signals, providing its genesis. Based on this, it is necessary to identify and study the mechanisms, supporting this process.

To date, the nature of the genesis of pacemaker electrical rhythms has not been fully investigated. In some types of spontaneously active smooth muscle, interstitial cells of Cajal (ICC) or ICC-like cells have been found. ICC-like cells (pacemaker cells of the gastrointestinal tract) are also found in the urethra, portal vein, urinary bladder, and ureter [57–60]. Moreover, the studies on ICC-like cells isolated from the urethra and urinary bladder showed their ability to generate spontaneous activity.

It has been experimentally shown that these cells are the main in generation and coordination of the muscular electrical activity, which creates peristalsis in the gastrointestinal tract. The existence of cells having complex geometry and terminal processes, such as the interstitial cells of Cajal, was also described in the myometrium [43]. Despite the nature of these cells has not been studied yet, we can not exclude their function similar to the interstitial cells of the gastrointestinal tract, providing generation and coordination of the contractile activity. In the recently presented data on the study of coordination of miometrial contraction, there is a conclusion about abovementioned process regulation by detected interstitial cells of Cajal in the uterus, action of which is similar to the hormonal "sensors" [61].

By morpho-histochemical studies, we have identified a great amount of round or oval atypical cells with high levels of Ca^{2+}-dependent acid phosphatase activity in longitudinal sections of the ovarian part of uterine horn in intact rats. We have also found such cells in the cervical end of uterine horn and uterine corpus, but the greatest accumulation of them was detected in the ovarian end of horn. Under the oxytocin action, a sharp increase in cytoplasmic enzyme activity of intensive-colored atypical cells was observed. These cells are colored so dark in the ovarian part of uterine horn that gives the locus of their location a completely blacked-out form, in contrast to the cervical part of the horn and uterine corpus. Thus, detection the greatest number of atypical cells in the ovarian end of horn can confirm the fact of the longest duration of electrical discharges in this area and the leading role of this particular locus in genesis and coordination of activities of the subsequent downstream active regions.

Acknowledgements

This study has been completely performed by the budget money of L.A. Orbeli Institute of Physiology, NAS of Armenia, Yerevan.

Author details

Naira G. Hunanyan[1,2*], Knarik V. Kazaryan[1] and Tatevik A. Piliposyan[1,3]

1 Department of Smooth Muscle Physiology, Orbeli Institute of Physiology, Yerevan, Armenia

2 University of Traditional Medicine, Yerevan, Armenia

3 Yerevan Haybusak University, Yerevan, Armenia

*Address all correspondence to: naira_hunanyan@yahoo.com

References

[1] Chen CJ, Chiang ST. Activities of uterine muscles from rats in late pregnancy. Proceedings of the National Science Council, Republic of China. Part B. 1989;**13**:42-46

[2] Garfield RE, Maner WL. Physiology and electrical activity of uterine contractions. Seminars in Cell & Developmental Biology. 2007;**18**:289-295

[3] Buhimschi CS, Saade GR, Buhimschi IA, Gokdeniz R, Boyle MB, Garfield RE. Effect of stimulatory and inhibitory drugs on uterine electrical activity measured noninvasively from the abdominal surface of pregnant rats. American Journal of Obstetrics and Gynecology. 2000;**183**:68-75

[4] Buhimschi C, Boyle MB, Garfield RE. Electrical activity of the human uterus during pregnancy as recorded from the abdominal surface. Obstetrics and Gynecology. 1997;**90**:102-111. Available from: https://pubmed.ncbi.nlm.nih.gov/9207823/

[5] Buhimschi C, Garfield RE. Uterine contractility as assessed by abdominal surface recording of Electromyographic activity in rats during pregnancy. American Journal of Obstetrics and Gynecology. 1996;**174**:744-753

[6] Kazaryan KV, Hunanyan NG, Akopyan RR. Electrophysiological characteristics of the rat uterine and adjacent horn areas. Russian Physiological Journal I.M. Sechenova. 2010;**96**:981-987 (in Russian)

[7] Maul H, Maner WL, Saade GR, Garfield RE. The physiology of uterine contractions. Clinics in Perinatology. 2003;**10**:665-676

[8] Rabotti C, Mischi M, van Laar I, Ou G, Bergmans J. Electrohysterogrhaphic analysis of uterine contraction propagation with labor progression: A preliminary study. In: Conf. Pros. IEEE Eng. Med. Biol. Soc. 2007. pp. 4315-4318

[9] Brar HS, Platt LD, De Vore GR. Qualitative assessment of material uterine and Fetal Umbilicat artery blood flow and resistance in Laboring patients by Doppler velocimetry. American Journal of Obstetrics and Gynecology. 1988;**158**:952-956

[10] Wray S. Hypoxia in the uterus. News in Physiological Sciences. 1994;**9**:88-92

[11] Cavaco-Goncalves S, Marques CC, Horta AE, Fique-roa JP. Increased cervical electrical activity during oestrus in Progestagen treated ewes: Possible role in sperm transport. Animal Reproduction Science. 2006;**93**:360-365

[12] Garfield RE, Bytautiene E, Vedernikov YP, Marshall JS, Romero R. Modulation of rat uterine contractility by mast cells and their mediators. American Journal of Obstetrics and Gynecology. 2000;**183**:118-125

[13] Joung RC, Schumann R, Zhang P. Nifedipine block of Capacitative calcium entry in cultured human uterine smooth-muscle. Journal of the Society for Gynecologic Investigation. 2001;**8**:210-215

[14] Marshall JM. Regulation of activity in uterine smooth muscle. Physiological Reviews. 1962;**42**:213-227

[15] Arthur P, Taggart MJ, Mitchell BF. Oxytocin and parturition: A role for increased myometrial calcium and calcium sensitization? Frontiers in Bioscience. 2007;**12**:619-633

[16] Phaneuf S, Europe-Finner GN, Carrasco MP, Hamilton CH, Lopez BA.

OxytocinSignalinginhumanmyometrium. Advances in Experimental Medicine and Biology. 1995;**395**:453-467

[17] Rabotti C, Mischi M. Propagation of electrical activity in uterine muscle during pregnancy. Acta Physiologica. 2015;**213**(2):406-416

[18] Meliksetyan IB. The identification of the Ca^{2+} dependent acid phosphates activity in cellular structures of the brain in rats. Morphology. 2007;**131**:77-80

[19] Chilingaryan AM. New Ca-adenosinetriphosphate method for revealing of Intraorganic microcirculating bed. NAS Armenia Reports. 1986;**82**:66-71

[20] Kao CY. Long-term observations of spontaneous electrical activity of the uterine smooth muscle. The American Journal of Physiology. 1959;**196**:343-350

[21] Mancinelli R, Guariglia L, Racanicchi C, Bertuzzi A, Salinari S, Vitelli R. In vitro motility of non-pregnant rat uterine horns. Quarterly Journal of Experimental Physiology. 1988;**73**:459-469

[22] De Lucas JJ, Rodriguez C, San AM, I., Ballesteros E. Measurement of spontaneous motility in vitro of the cervical ring in sheep. Revista Española de Fisiología. 1988;**44**:69-72

[23] Kao CY. In: Wynn RM, editor. Biology of the Uterus. New York: Plenum Press; 1977. pp. 423-496

[24] Bengtsson B. Factors of importance for regulation of uterine contractive activity. Acta Obstetricia et Gynecologica Scandinavica. Supplement. 1982;**108**:13-16

[25] Bortoff A. Myogenic control of intestinal motility. Physiological Reviews. 1976;**56**:418-435

[26] Hamilton WF. Circulation. In: Handbook of Physiology Series, Section 2. Vol. 2. Bethseda: American Physiological Society; 1963

[27] Even MD, Laughlin MH, Krause GF, vom Saal FS. Differences in blood flow to uterine segments and placentae in relation to sex, intrauterine location and side in pregnant rats. Journal of Reproduction and Fertility. 1994;**102**:245-252

[28] Harrison N, Larcombe-McDouall JB, Earley L, Wray S. An in vivo study of the effects of ischaemia on uterine contraction, intracellular pH and metabolites in the rat. The Journal of Physiology. 1994;**476**:349-354

[29] Wray S, Duggins K, Iles R, Nyman L, Osman V. The effect of metabolic inhibition and intracellular pH on rat uterine force production. Experimental Physiology. 1992;**77**:307-319

[30] Larcombe-McDouall JB, Harrison N, Wray S. The in vivo relationship between blood flow, contractions, pH and metabolites in the rat uterus. Pflügers Archiv European Journal of Physiology. 1998;**435**:810-817

[31] Maynard R, Downes N. Anatomy and histology of the laboratory rat. In: Toxicology and Biomedical Research. 1st ed. London, UK: Academic Press; 2019. p. 378

[32] Mc LA, Michie D. Control of pre-natal growth in mammals. Nature. 1960;**187**:363-365

[33] Van Saal F, Dhar M. Blood flow in uterine loop artery and loop vein is bidirectional in the mouse: Implications for transport of steroids between Feluses. Physiology and Behavior. 1992;**52**:163-171

[34] Crane LH, Martin L. Pace-maker activity in the myometrium of the

oestrous rat: In vivo studies using video-laparoscopy. Reprod. Fertil. Develop. 1991;**3**:519-527

[35] Hertelendy F, Zakar T. Prostaglandins and the myometrium and cervix. Prostaglandins, Leukotrienes, and Essential Fatty Acids. 2004;**70**:207-222

[36] Pajntar M, Verdenik I, Pusenjak S, Rudel D, Leskosek B. Activity of smooth muscles in human cervix and uterus. European Journal of Obstetrics, Gynecology, and Reproductive Biology. 1998;**79**:199-204

[37] Pajntar M, Leskosek B, Rudel D, Verdenik I. Contribution of cervical smooth muscle activity to the duration of latent and active phases of labour. BJOG: An International Journal of Obstetrics and Gynaecology. 2001;**108**:533-538

[38] Ferreira JJ, Butler A, Stewart R, Gonzalez-Cota AL, Lybaert P, Amazu C, et al. Oxytocin can regulate myometrial smooth muscle excitability by inhibiting the Na+−activated K+ channel, Slo 2.1. The Journal of Physiology. 2019;**597**(1):137-149

[39] Garfield RE, Beier S. Increased myometrial responsiveness to oxytocin during term and preterm labor. American Journal of Obstetrics and Gynecology. 1989;**161**:454-461

[40] Mironneau J. Effect of Oxytocine on ionic currents underlying rhythmic activity and contraction in uterine smooth muscle. Pflügers Archiv. 1976;**363**:113-118

[41] Hashitani H, Garcia-London AP, Hirst GDS, Edwards FR. Atypical slow waves generated in gastric corpus provide dominant pacemaker activity in Guinea-pig stomach. The Journal of Physiology. 2005;**569**:459-465

[42] Sergeant GP, Hollywood MA, McCloskey KD, Thornbu-ry KD, McHale NG. Specialized Pacemaking cells in the rabbit urethra. The Journal of Physiology. 2000;**526**:359-366

[43] Duquette RA, Shmygol A, Vaillant C, Mobasheri A, Pope M, Burdyga T, et al. Vimentin-positive, c-KJT-negative interstitial cells in human and rat uterus: A role in Pacemaking? Biology of Reproduction. 2005;**72**:276-283

[44] Dodds K, Travis L, Beckett E, Spencer N. Identification of a novel distension-evoked motility pattern in the mouse uterus. American Journal of Physiology. Regulatory, Integrative and Comparative Physiology. 2021;**321**(3):R317-R327

[45] Edwards D, Good DM, Granger SE, Hollingsworth M, Robson A, Small RC, et al. The Spasmogenic action of oxytocin in the rat uterus-comparison with other agents. British Journal of Pharmacology. 1986;**88**:899-908

[46] Hunanyan NG, Kazaryan KV, Tirayan AS. The study of spontaneous electrical activity registered from different regions of the rat uterus. In: The Modern Problems of Integrative Activity and Plasticity of the Nervous System. Materials of Conference. 2009. pp. 306-311 (in Russian)

[47] Kazaryan KV, Hunanyan NG, Meliksetyan IB, Savayan AA. The effect of oxytocin on spontaneous activity of uterine horn in rats. NAS Armenia Reports. 2010;**110**:397-406 (in Russian)

[48] Kazaryan KV, Hunanyan NG, Mkrtchyan AV. Spontaneous electrical activity of the fallopian tubes in rats. Journal of Evolutionary Biochemistry and Physiology. 2020;**56**(3):245-250 (in Russian)

[49] Norwitz ER, Robinson J, Challis JRG. The control of labor. The New England Journal of Medicine. 1999;**341**:660-666

[50] Wolfs GMJA, van Leeuwen M. Eiectromyographic observations on the human uterus during labour. Acta Obstetricia et Gynecologica Scandinavica Supplement. 1979;**90**:1-61

[51] Garfield RE, Blennerhassett MG, Miller SM. Oxford Rewiews of reproductive biology. In: Clarke JR, editor. Control of Myometrial Contractility: Role and Regulation of Gap Junctions. Vol. 10. Oxford: Oxford University Press; 1988. pp. 436-490

[52] Orlov RC, Izakov VY, Ketkin AT, Plechanov IG. Regulating Mechanisms of Smooth Muscle Cells and Myocard. Leningrad, Russia: Nauka; 1971

[53] Shuba MF. The effect of sodium-free and potassium-free solution, ionic current inhibitors and Ouabain on electrophysiological properties of smooth muscle of Guinea-pig ureter. The Journal of Physiology. 1977;**264**:837-851

[54] Garfield RE, Daniel EE, Dukes M, Fitzgerald JD. Changes of gap junction in myometrium of Guinea-pig at parturition and abortion. Canadian Journal of Physiology and Pharmacology. 1982;**60**:335-341

[55] Olah KS. Changes in cervical Electromyographic activity and their correlation with the cervical response to myometrial activity during labour. European Journal of Obstetrics, Gynecology, and Reproductive Biology. 1994;**57**:157-159

[56] Hunanyan NG. The effect of Anemia on electrical activity of ovarian ending of uterine horn in rats. Medical Science of Armenia. 2012;**52**:33-39 (in Russian)

[57] Klemm MF, Exintaris B, Lang RJ. Identification of the cells underlying pacemaker activity in the Guinea-pig upper urinary tract. The Journal of Physiology. 1999;**519**:867-884

[58] McCloskey KD, Gurne YAM. Kit positive cells in the Guinea-pig bladder. The Journal of Urology. 2002;**168**:832-836

[59] Pucovsky V, Moss RF, Bolton TB. Non-contractile cells with thin processes resembling interstitial cells of Cajal found in the wall of Guinea-pig mesenteric arteries. The Journal of Physiology. 2003;**552**:119-133

[60] Spencer NJ, Greenwood IA. Characterization of properties underlying rhythmicity in mouse portal vien. Autonomic Neuroscience. 2003;**104**:73-82

[61] Huchings G, Williams O, Cretoiu D, Ciontea SM. Myometrial cells and the coordination of myometrial contractility. Journal of Cellular and Molecular Medicine. 2009;**13**:4268-4282